LOFT
LIVING

LOFT LIVING

Recycling Warehouse Space for Residential Use

Kingsley C. Fairbridge
Harvey-Jane Kowal

Saturday Review Press/E. P. Dutton & Co., Inc.

All photographs and illustrations are by Kingsley C. Fairbridge except the following: pp. 9 (top), 58—David Friedman; p. 89—Richard Haber; pp. 95, 103, 107—Harvey-Jane Kowal; pp. 104–105—Films at Cost; and p. 125—Ron Lieberman.

First Edition
10 9 8 7 6 5 4 3 2 1

Published simultaneously in Canada
by Clarke, Irwin & Company Limited, Toronto and Vancouver

ISBN: 0–8415–0426–1 (cloth)
0–8415–0434–2 (paper)

Library of Congress Cataloging in Publication Data
Fairbridge, Kingsley C
Loft living.
Includes index.
1. Dwellings. 2. Buildings—Repair and reconstruction. 3. Warehouses. I. Kowal, Harvey-Jane, joint author. II. Title.
TH4812.F34 1976 690.8 75-31852

For Diane and Richard

Acknowledgments

Thanks to the following for the help they gave us: Bruce Baratz, Marcia and Alan Docter, David Friedman, Kathleen Kelly-Mazza, Trusha Kushner, Jerry Lieberman, Ron Lieberman, Don Lubov, Gypsy Lujan, Richard V. Mazza, Gene Mekler, Gary Rieveschl, and Wendy Teitelbaum. Special thanks to Sydney Wolfe Cohen.

Contents

PART I
PREPARATIONS FOR LOFT LIVING

.1. Introduction

In the past 15 years working and living in lofts has become a way of life for an increasing number of people. Lofts used to be restricted to a specialized segment of the population—painters, sculptors, actors, and dancers in need of large work space. When artists first rented large commercial space to work in, their economic status generally precluded renting an apartment to live in and maintaining a studio as well. Thus they began to use their work space for living.

Once these early loft dwellers got used to living in open space, they became spoiled and no longer wanted to live in apartments. And in the early 1960s there was an increasing demand on the part of young people in general for alternative life styles.

In addition, economic factors have made available large amounts of commercial space. Because of escalating taxes and rising trucking, labor, and utility costs, many industries that traditionally rented such space either have gone out of business or have moved out to suburban industrial zones, thereby leaving empty many factories, warehouses, and old brownstones.

The flight of the middle class to the suburbs has slowed. Many people who might have thought about moving out of urban areas, but who now cannot or even don't want to, are thinking of alternative ways of living in cities.

In addition to artists, a large part of the urban population has become dissatisfied with the space limitations of living in conventional-sized apartments. New construction costs are so high that landlords are finding it increasingly difficult to build large-sized apartments, except for the ultrarich. Most new construction is lighter and flimsier and therefore less soundproof than older, thicker-walled structures. People looking for refuge from the noise of city streets and concentration of people in residential "neighborhoods" are simply seeking a measure of elbow room, quiet, and isolation. Since few if any new buildings meet these requirements of size and sturdiness, city dwellers are looking toward the reutilization of older structures to satisfy their needs.

(*Above*) The openness of the average loft allows for almost limitless possibilities in terms of layout for living and working.

(*Right*) Most lofts are in old manufacturing and commercial neighborhoods.

At one time it was impossible for anyone to rent commercial space for living because of strict enforcement of zoning regulations against such a practice. Even today much of this space in the pure legal sense is rented only for commercial use as "artists' work space." Due to economic necessity, landlords are changing. They are less strict in their interpretation of "living" or "working" space and look away to a great extent in order to gain income from otherwise unrentable space. Also, many city governments are making living in certain commercial areas quasi-legal to lessen the housing shortage and to gain lost tax revenue. In New York City the area known as SoHo—a section of the city south of Houston Street, bounded by West Houston and Canal streets—was rezoned in 1971 by the City Planning Commission to allow artists to live legally in certain factory buildings.

Though the loft-living phenomenon probably began in New York City as a result of the large concentration of light industry, it has spread to all urban areas of the country where there are suitable buildings. In

almost any city—or any town—where you can find a house, you can find commercial space. (Before signing a lease on commercial space, you should check local zoning laws carefully to see whether it is legal for you to live there.) Generally this space is in the downtown area of cities, where most industry was located.

Technically, the term *loft* refers to one of the nonpartitioned upper floors of a warehouse or business building. Today the term is applied more broadly to any kind of factory space or warehouse—in fact, to recycled buildings of any size, shape, and structure, such as supermarkets, garages, and railway stations—and is certainly not restricted to upper floors. The terminology used to refer to this type of space varies from coast to coast. On the East Coast these spaces are usually called lofts; on the West Coast they are more often referred to as warehouses. We will use the term *loft* throughout this book, but we will deal with many of the various configurations, including brownstones (row houses made from reddish-brown sandstone) gutted out as floor-throughs—that is, unpartitioned floors. (Very often brownstones have gone through several phases, starting as residential townhouses and then, as neighborhoods changed, being converted to light industrial use. In the Northeast at the turn of the century brownstones housed such industries as dressmaking and tinsmithing shops and sometimes even dead storage.)

The "average" loft is often at least 25′ × 100′, but much larger space can be found. Most brownstones are relatively small—approximately 20′ × 60′—and often seen in two-floor "duplex" configurations. Many brownstones are walkups of four flights or more. You can often find ten-story buildings with automatic elevators—or very old wood-frame buildings that still have rope-drawn manual elevators.

Whatever the size or shape of a loft, whatever you choose to call it, this book will deal with the basic techniques needed to turn any raw, previously commercial space into viable living and working areas.

.2.
Finding and Checking Out Prospective Lofts

Some people need loft space to work in; others are just looking for a place to live. Still others are looking for space to live *and* work in. In some parts of the country people seek out and rent commercial space simply because it is still extremely cheap. In Project Artaud, a warehouse community in San Francisco, space goes for as little as ten cents a square foot. Another reason for renting commercial space for living is the communal spirit. There are a number of commune-type loft buildings across the country where the various tenants share kitchen, bathroom, and living quarters. Actors and dancers are attracted to warehouse and factory buildings because they need large, unobstructed space in which to rehearse or conduct workshops. And handicapped persons in wheelchairs are finding loft space ideal for living. Because of the openness and lack of partitions, lofts can be designed to fit the specific needs and requirements of people confined to wheelchairs.

Whatever your reason for seeking loft space, you must consider your particular requirements in order to find the right loft for you. A metal sculptor might be better in a ground-floor loft with a solid concrete floor and a loading platform; a writer might be happier with the top floor of a small building on a quiet street. Before locating your loft, you should give careful thought to what you want to do in the space as well as to what things you must have and where you are willing to compromise. If you want to open a photography studio, a good water supply is essential, as well as high ceilings and a central location convenient for clients. A musician who wants to play his instrument at any hour of the day or night must consider a suitably isolated location. A musician who wishes to live in a multifamily dwelling will have to consider the labor and financial costs involved in soundproofing (which will be dealt with later in detail) in order to co-exist with the other tenants in the building. A light, airy place would be of prime consideration if you wanted to grow trees and hanging plants. You might be

Some people use lofts just for working.

This loft is used exclusively for living.

A typical working and living loft.

willing to live in a more remote area in exchange for a top-floor loft with many skylights.

Financial resources are of paramount importance: not only how much rent you can afford to pay but also how much reserve cash you have or are willing to put into renovation. You can rent raw space and put in work little by little as cash flow allows. You do not necessarily have to do all the work immediately. This is a question of what you consider to be minimal living conditions and how much mess you are willing to tolerate over a period of time. Most people like to have at least a shower, toilet, and sink. If you can budget for it and wish to renovate a completely raw loft, it's best to live elsewhere while doing the basic work.

In the course of renting a loft that has been lived in previously, fixture fees or "key money" is often involved. You might find it preferable to rent a completely raw loft rather than pay someone else for work that's been done before you. Or, if the space is desirable but the existing fixtures are inferior or not to your liking, you might be willing to pay the fee anyway and then remodel to suit your needs. This is a matter of priorities. What it often comes down to is either taking raw space directly from the landlord, which is the cheapest way initially, or spending $1,000 to $10,000 and upward for fixtures and moving in.

Once you have established your priorities and possible areas of compromise, you are ready to begin the search. There are many different ways to go about finding a loft, depending on which city you live in. One of the easiest and most direct ways is to walk or drive around an area and look for signs—"Loft for Rent," "Space for Rent," and so on. Often the superintendent or elevator operator will show you vacant floors. Then, if you are interested, you can contact the landlord or agent. Another way is to speak with people already living in commercial areas. Getting the word out is often very helpful. People living in lofts are usually the first to hear about other lofts that are available.

If your city has rental agencies that deal in commercial space, you should register with them. You'll also find that landlords in many cities advertise in the local newspapers, as do loft dwellers interested in selling fixtures and subleasing their lofts.

It's best to survey the area you're interested in—and not just once. Check it at various times during the day. Remember, you are dealing mostly with commercial areas that can be quiet and deserted at 1:00 P.M. but in full swing at 1:00 A.M. In the wholesale florist district of New York City, for example, refrigerator trucks deliver flowers between 11:00 P.M. and 5:00 or 6:00 A.M., and the shops are open from about 5:00 A.M. to 12:00 P.M. It can be very difficult to sleep through this noise. If you're a carpenter who works late into the night, this may not bother you. If you have to get up at 7:30 to go to work, this could be an intolerable condition. It's best to know what you're getting into *before* you sign the lease.

CHECKING OUT PROSPECTIVE LOFTS

When you've located a loft that appeals to you in terms of location, size, and personal requirements, it's essential to check out some of the basic systems. Do not rush into the first loft you see. Time spent with a flashlight and tape measure on the floor checking such things as gas, water supply, and electrical services, as well as the condition of any heating system in the building, can save you heartache later on. It is best to check the basement of the building for the source of these essential services. Sometimes spectacular space may not have some of the basic services you want or expect. There may be no gas supply, poor or antiquated electrical service, or severely corroded or blocked water lines and drains. At this point you will have to be prepared to either invest heavily in basics or pass on to the next loft.

The following is a list of basic points you can check for without professional help, though on some of these, depending on your plans for the space, professional advice could help you make a more knowledgeable decision.

Plumbing

Check for a functioning toilet *within* the confines of the space you are interested in renting. If possible, avoid space where the only existing toilet is in the hallway. Bringing a toilet into this type of space can entail a large financial outlay for the heavy work of punching a hole through your floor and running pipe across a lower floor's ceiling. Also, most landlords prefer to have this kind of work done by licensed professional plumbers because of the extensive paperwork involved in filing building plans and the possibility of receiving violations and fines.

Some old factory floors have twin toilets (a requirement when men and women work on the same floor). This arrangement can be very useful to a loft dweller in that one of the toilets can be removed and a shower stall or bathtub installed and drained into the old toilet opening.

Water pressure and drainage should also be checked. The easiest way to test for good pressure and drainage is to open the tap on any existing sink or tub. If the water comes out rusty, fails to go down the drain, or drains slowly, you have occluded pipes and blocked drains. If there is no sink or tub in which to check pressure, flush the toilet and see how long it takes to fill up—60 to 90 seconds is OK; anything longer indicates a potential pressure problem. Next, check the main riser—the building supply line—coming up from the basement. It should be at least ¾″ in diameter. You should also check the reduction point at which the main riser branches off to your floor. You may have a ½″ galvanized pipe that is heavily rusted, thereby ruining your pressure. Watch out for ½″ galvanized supplies in general—they usually rust up

An existing twin toilet arrangement can be utilized for a new drainage hookup for shower or tub.

and tend to block more quickly. Brass or copper tubing is the best material, as it never rusts or corrodes. In some cases landlords may be induced to correct inadequate or antiquated plumbing. For example, if your water pressure is extremely poor or almost nonexistent, the landlord might pay to have the work done or give you a break on the first month's rent and have you do the work.

Remember that water pressure is also contingent upon usage elsewhere in the building (and in some cases in the street). Heavy water-using tenants and roof-mounted water-fed air conditioners can substantially reduce the water pressure available to you.

Electricity

Unless you know something about electricity, it might be useful to take along a knowledgeable friend or an electrician to check it out.

Sometimes you will find high-amperage electrical service (twin 60-amp lines or more) on industrial floors because large machinery and air conditioners were used by the previous commercial tenant. If this is the case, you are lucky because you will probably never need all the electricity that the previous tenant used. Look for modern circuit-breaker boxes and, if possible, remove the cover plate (usually attached by just four screws) and check wire size of supply. Wire size is a good indication of the amount of power available on your floor, or in your space. The thicker the wires, the greater the amount of power available. For example, two No. 10 wires (each about ⅛″ thick) mean that you will have only a twin 30-amp capability, enough to run a minimal amount of electrical equipment—such as refrigerator, lights, a stereo set, and toaster—but definitely not enough for large air conditioners, paint spray compressors, steam machines, saunas, or any other heavy-duty electrical devices. If you have a service-entrance box with two or three No. 6 wires (each about as thick as a pencil), you will have plenty of current for whatever you may have in mind. Remember that wire sizes are inversely related to their thickness. No. 16 (commonly seen on lamps, small appliances, and lightweight extension cords) is fairly thin. No. 2 wire is very thick and capable of much higher load capacity.

This is an example of minimal wiring. There is not enough amperage in this box to run more than a minimum number of household appliances and certainly would be insufficient for large-scale air conditioning.

Be careful when opening up breaker boxes to check for adequate wiring—even a 30-amp load can kill you if you ground yourself out while touching a terminal. *Look, but don't touch!* If there is any doubt, or if you want specific information about your setup, get an electrician to make an inspection of the wiring. His inspection fee may save you money in the long run. If you have to run new wiring up several flights for your floor, it might cost well over $1,000, depending on the size wire you'll need, the length of the run, and the price of copper on the day the wire is bought. (The demand for copper shell casings for the war in Indochina skyrocketed the price of copper many years ago; and with current inflation and new awareness on the part of copper-exporting countries, wire has become terrifically costly.)

A nice minimum supply would be two No. 6 wires. Avoid a loft with just a couple of low-amperage (15- or 12-amp) screw-in type fuses, or be prepared for major electrical supply work. If you are planning to rent such a loft, it is imperative to check that the building itself has an adequate supply. Most utilities will bring the larger supply in from the street free, but it is quite expensive to have the larger supply switch installed for the whole building. Some landlords will undertake such work, but be prepared for a major expenditure and plenty of hassles with the utility company and its inspectors.

Be sure also to check whether you have AC or DC electrical current. If you have DC, you can either buy transformers for all your appliances—at great expense—or have the service changed to AC by the electric company. Again, the utility will bring the service into the building free, but you will have to rewire the rest of the building at some cost. A landlord may be persuaded to incur this expense, as the modernization improves the value of his building.

Check for separate electrical meters for each floor. Very often in old buildings two or three floors are connected to a single meter. To have your floor on a separate meter could entail running new electrical tubing and wire down six or seven flights to the basement, which can definitely be costly.

If you have adequate service to the floor, don't be worried if there are few or no outlets around the walls—these can be readily installed along with any changes you may want in your overhead lighting.

Gas

Although the supply of natural gas is very low right now, and there are few new installations of gas service, many old buildings have existing supply pipes coming in from the street that can be used for auxiliary heating, cooking, and hot water. Chefs usually concede that gas is superior to electricity for cooking, and if it's available, it's nice to have. Also, in most parts of the country gas is still less expensive for heating and cooking than electricity.

If gas service is available, make sure there is a meter for your floor in the basement. Some utilities are not installing new meters, so check with your local utility company if you are planning to revitalize an old gas line. Always ask for a residential rate; even though the address may be considered subject to higher commercial rates, you are entitled to the lower rate as a residential tenant. (This applies to electrical service as well.)

Heating

Is there existing building-supplied heat? Is the furnace steam- or gas-fed? If heating is provided, check on the hours it operates. Often commercial buildings have heat only during business hours, usually from nine until five Monday through Saturday. Some landlords can be persuaded to reset the timing equipment on their furnaces to provide heat seven days a week on an extended daily schedule. Since the landlord has to pay the fuel oil costs, he may not like the idea, but it's worth a try. Also, keep in mind that with the cost of installing supplemental heating units, in the long run it might be cheaper to offer the landlord an increased rent for the winter months to cover his additional fuel costs. If the building is still primarily commercial and the landlord doesn't want to incur this expense, you will have to install auxiliary heating.

If you are lucky enough to have a gas supply line, space heaters or ceiling-hung blowers can be installed easily and are efficient and economical. Take into consideration the extra cost of heating in addition to your regular utility bill. If you do not have gas, your only recourse will be electrical baseboard or unit heaters, for which you will need a large electrical supply and a correspondingly large income to pay for it.

The best setup to have is a steam system that is maintained by the landlord and left on all the time.

In some buildings there is no existing heat at all, in which case you will have to pay for your own heat; you should take this into account when negotiating your rent. In this situation at least you are master of your own destiny in governing your heat. You can turn it on when it's cold outside and off when you're not home, thereby saving energy and money. Also, if it's cold in April or May, you can still have heat, whereas many buildings with central heating have a turnoff date of, say, March 21, regardless of the temperature outside.

Structural Aspects

Floors. Ideally, you would like to walk into an old factory loft and find good maple (or any other hardwood) floors ready to be sanded and finished. Fortunately, this is often the case. With proper sanding and refinishing, you can often get a spectacular expanse of wood floor.

Concrete floors are also quite common, but they are not really suitable for comfortable living. Depending on your needs, however, a concrete floor could be useful. If you plan to do some heavy work, such as welding metal sculptures, you might want to retain part of the concrete floor for your work area and cover the living areas with hardwood or foam-backed wall-to-wall carpeting. Putting down large amounts of carpeting or wood flooring can be quite costly, so you might want to look for a place that has wood floors.

In older wood-frame buildings check for floor sag. This is the result of aging floor beams, and whether you can live with it is really a matter of taste. It is generally very expensive to correct this condition, which entails major structural work—gutting out the entire building and replacing the old wood beams. A small (1″ or 2″) depression can be leveled out with shims—material inserted under another material to raise, level, or bring it flush—and refloored.

Old floors can be brought back to life with sanding and refinishing. The luster of the natural wood of a floor that had been neglected and abused for years (*above*) can be restored by sanding and polyurethaning.

Severely worn wide-plank floors are usually impossible to refinish because of their large knots; it is usually best to carpet them, using a heavy double-foam underpad. Or they can be livened up with a couple of thick coats of good-quality deck and porch paint. If money is no object, a new floor can be laid over the old one. New flooring ranges from $1 to $2 per square foot.

Walls. Check the general condition of any existing walls. Is there water corrosion, falling plaster, severe plaster decay? If you are in a brick building and the walls are plastered over, you may want to strip them down to raw brick.

If the space is subdivided and you are interested in opening it up, check with an engineer as to load-bearing and nonload-bearing walls and the feasibility of removing them. It's often possible to open up space by removing unnecessary partitions. Keep in mind that you may need a buildings department permit to undertake such modifications.

Ceilings. Inspect ceilings for leakage. If you are on the top floor, make sure the landlord will take responsibility for any necessary repairs to the roof, since a major roofing job can be very expensive.

Windows. You may come across a number of different types of windows when looking at lofts—double-hung sash windows, pivoting picture or plate-glass windows, or casement windows. Whatever type you encounter, make sure they open easily and shut tightly. If they're double-hung sash windows, do they slide smoothly? Watch out for windows that don't open easily because they are jammed with paint. Check frames for tightness or any openings between frame and wall. Is the entire window area well sealed? Is the glass in good condition? Are the windows loose and rattly?

Repairing and replacing windows can be an expensive undertaking. Landlords will usually replace broken glass; in fact, they very often have insurance policies that cover all the windows in the building. However, replacing complete window frames will often have to be done at your expense.

If you have young children, keep in mind that lofts with ceiling-to-floor pivoting windows have very low sills and can be dangerous.

Skylights. Skylights can be very spectacular and are showpieces in many lofts. With the big boom in growing indoor plants, loft space with skylights tends to rent at a premium. You can even put them in if there aren't any. One of the disadvantages of skylights is that they make a loft hotter in the summer, and if you're going to have air conditioning, you'll need a more powerful unit.

If the loft has a skylight, check carefully for signs of water damage on the surrounding ceiling. Skylights require first-class maintenance to keep water from coming in, which usually means redoing flashing on the roof fairly often. Check glass for cracks, as replacing it can be a major expense. As with windows, landlords should take care of this, but many will not.

If sections in the skylight can be opened, test the action of the mechanisms to see if they work. If they do not open and close properly, you can have terrible problems with water damage, as well as heat loss in the winter.

Elevators. Many factory and loft buildings are equipped with elevators. These range from hand-drawn, rope-fed types to the most modern, self-service units. Sometimes there are even elevator-operator units. These generally have a man on duty only during business hours, five or six days a week. If this is the case, find out from the landlord or superintendent if you can get a key to use the elevator the rest of the time. If this isn't possible, be prepared to walk up flights of stairs nights and on weekends. A couple we know bought into a building that has a hand-operated elevator and no elevator operator. At the time they took the loft they never foresaw any problem living on the eighth floor of a building with such an elevator. They learned quickly. After the first few nights of being awakened at 3:00 A.M. to bring the elevator down for their ninth-floor neighbors who didn't feel like walking up, they got into walking up themselves. If you don't like to walk up flights and flights of stairs, stay clear of such buildings or rent only lower floors. Slowly but surely landlords are installing self-service elevators, and you may be lucky enough to find a building with one.

Miscellaneous. A note for specialized artists and craftsmen. If you are going to be working on a large scale, make sure that there is a doorway or window large enough to get your pieces through! This may seem elementary, but it should be mentioned; sculptors also need big doors, and they should check the floor's load rating and capability as well as the availability of heavy electricity for arc welders. Potters will want good water supplies as well as good drainage, as the waste from most pottery sinks builds up quickly and needs to be cleaned out frequently. Most potters also require a good electrical supply on the floor for their kilns. If you need heavy electricity, try to get a floor that was once a factory, since it will probably already have a big panel board or breaker box. Otherwise you will have to spend quite a lot of money to bring up heavy electrical service from the basement to your floor.

Skylights make
growing houseplants easy.

.3.
Leases, Fixture Fees, and Finances

LEASES

Depending on what part of the country you live in, you may or may not be able to get a lease for the space you wish to rent. Having a lease protects your investment of time and money from the capriciousness of your landlord. Try to get as long a lease as possible, with a sublease clause to give you some flexibility.

Most loft space is considered to be commercial in nature and is therefore not covered by any form of rent control other than the traditional market pressures of supply and demand. Since the space is commercial, from a strictly legal point of view it should be used for commercial purposes only—that is, as an art studio, photography studio, and so on. However, as mentioned earlier, most landlords are anxious to rent the space and are relatively lenient in their renting practices. Often they are not interested in knowing exactly why you are renting the loft or whether you are going to live there. We have generally found this to be the case coast to coast. Under these conditions, if you can get a lease, it will be strictly commercial. If you wish to be up front with the landlord, you might be able to persuade him to write a residential lease, providing zoning regulations in your city permit.

A friend of ours in Cincinnati went so far as to have the entire downtown area rezoned so that he could live legally in a commercial building. He did this without the aid of a lawyer by simply presenting himself to the city council and asking them how he could have his building rezoned residential. Basically, all he had to do was meet certain city building codes, such as using a licensed plumber for his plumbing installations and installing a fireproof floor. His landlord would have given him a lease regardless, but he decided to protect his investment.

A residential lease can be useful for obtaining residential rates from gas and electric utilities as well as for legitimizing your living. Residential leases may also help you avoid paying commercial-occupancy taxes.

If a lease is a possibility, it is preferable to get a relatively long one (three to five years at least) with a sublease clause. This will enable you to sell your fixtures and give any possible subtenant a fixed rate or at least a controlled escalation of the rent for the balance of the lease.

There can be a great deal of flexibility in negotiating a lease because of the many different factors that come into play. Rent can be fixed over a short period of time (one to two years); or you can write a lease where the rent is raised periodically over a longer period (three to five years). Then again, you can sometimes find ten-year leases, but these are mostly net leases that entail renting the entire building and being responsible for ten years' worth of rent and in some cases the taxes and operating expenses for the building.

Landlords will in most cases offer you *their* lease to sign—read it very carefully, as many landlords add or delete clauses to their advantage. Leases in general give most rights to the landlord and few if any to the tenant, except for the security of being assured a specific amount of space for the specified amount of time for an agreed-upon amount of rent. Be careful! If you don't read your lease, you may find yourself paying a share of the real estate tax, water tax, sewerage tax, and others, in addition to a yearly escalation in your rent. Watch out for this—it is a deceptive practice. Escalating rent clauses are supposed to cover potential future increases in the cost of fuel oil and taxes. You should not pay a share of building taxes as well as an escalating rent.

Once you have studied your lease, make a list of the points you think are negotiable. If the landlord is asking for big yearly increases in rent and you think you can negotiate for a lower rent, point out all the weaknesses of the space—such as bad plumbing, poor condition of walls, or (if you have to provide your own heat) the projected cost of utilities. If the space is so raw that you will have to do major demolition work and subsequent construction (these things often take several months) or install basic life-support facilities such as toilet and shower, try to get a rent concession—that is, the first two or three months free or at a greatly reduced rate. Or, if the building needs a new hot-water riser and you are handy enough to do it, you might offer your services to the landlord in exchange for a rent abatement. You might also try to get a clause added to your lease giving you an option to renew at a specified increase, thereby saving you from having to haggle all over again when your present lease runs out.

It is good to establish clearly in the lease who is responsible for basic repairs to your space *before* you move in (or within a specified period after you move in) and what, if any, future repairs and maintenance the landlord will take care of. Get this in writing, as many landlords are notorious for making verbal commitments to fix things and never following through. The agreement should cover such basics as the roof, skylights, window glass, main drainage, and in some cases the cold-water supply to your floor.

In negotiating a new lease last year, we did some bargaining and countering with our landlord. At the expiration of our old lease we were paying $325 a month. We asked for a two-year lease at $350 a month or a five-year lease at $375 a month. The landlord didn't like our terms and asked for $400 for a five-year lease, which we found unacceptable. We then countered with a five-year program, starting at $350 the first year, $375 the second, $400 the next two years, and $425 the last year. We finally settled on this because it assured us and the landlord a controlled amount for five years. We also managed to write into the lease a clause stating that our landlord would repair our roof and waterproof a leaky exterior wall that was ruining our nicely stripped brick wall—maintenance long overdue—within six months of signing the new lease. It took him the full six months to get around to doing the work, and he did so only because the lease stipulated that if the work was not done within the period specified we would take care of the work and charge it against our rent.

A month's rent plus one month's security is the standard amount the landlord gets upon signing a lease. Avoid paying more than one month's security. Make sure that the lease specifies that the landlord has your security in his possession. In some states the law requires that the landlord deposit security in an interest-bearing savings account. We gave our landlord a month's security, which was mentioned on the bottom of our lease. About five months into the new lease he decided that he wanted another month's security. However, the lease was already in effect and we told him to read it—it was too late. He never bothered us about this again.

If you can't get a lease, see if you can get a written agreement from the landlord or renting agent entitling you to stay in the space for a specified amount of time, thereby justifying and protecting your investment. Remember that a written piece of paper, even if not a lease, can mean a lot more in a courtroom than a handshake with the landlord, no matter how honorable a person he may be.

FIXTURE FEES AND FINANCES

Raw space is nice if you can find it and if you have the resources and know-how to fix it up exactly to your specifications. However, many lofts have some existing fixtures put in by a previous tenant. In these cases fixture fees are often involved. The practice of selling fixtures is technically illegal, because most leases specify that "fixtures" permanently installed (that is, anything bolted or secured to the walls or floors, such as built-in furniture, sink, tubs—anything that cannot be lifted up and carried away) revert to the possession of the landlord at the expiration of the lease. However, the practice continues as an equitable method of compensating and protecting a tenant's investment of labor and money. In some instances a fixture fee may be nothing more than thinly disguised key money—that is, paying for the right to rent a spectacular view or large space with little or no actual fixtures involved.

What is a "just" or equitable price for fixtures? People calculate the value of their work in different ways. Some people include the value of their labor plus the cost of the materials and appliances; others do not include their labor and depreciate the value of appliances over the period of time they use them. Still others determine their prices on the basis of what the market will bear, or what similar-sized or similarly located lofts are bringing in. We cannot begin to give representational prices, as they vary widely within cities and across the country and range from $100 to as high as $10,000 or $15,000. These prices need not reflect the quality and type of work done. What may seem outrageous to some will seem fair to others. Keep in mind that a fixture fee is only one of many factors to be considered in the decision to rent loft space.

When taking over a place where you are paying for fixtures, make sure that the lease has a sublet clause and, if possible, meet with the landlord before you turn over the money. We know quite a few people who have been taken advantage of in such situations. Recently one of our neighbors sold his loft fixtures for $3,000, received the check, and left for parts unknown. The new tenant, who had never spoken with the landlord, found himself faced with eviction unless he agreed to pay a hefty rent increase. So insist on meeting with the landlord and having the lease signed over to you. This guarantees you the same financial terms for the balance of the previous lease.

There are some definite advantages to buying existing equipment and fixtures, because you can simply move in and start living and working. Or, if there is a minimal amount of preexisting work, you can have a basis from which to begin. If you are planning to redo everything, you may not want to incur the cost of paying for work you are going to rip out. This is a waste of money (and time); but again, if the space is really impressive, it may be worth it to pay the fee just to have it. This is a decision you alone can make.

Money is the single most important factor in the planning and execution of any loft project. Planning ahead on money matters can save you from mental distress and disaster later on. Because prices for materials are constantly rising and unexpected expenses always seem to crop up, the usual rule of thumb is to figure out your rough expenses and then double that figure. There never seems to be enough money to do all the things you would like to do. It depends on your own financial situation.

It's good to be able to start a project with at least a couple of thousand dollars in hand, but many people must simply move in and do work as weekly or monthly income allows. This process can be slow and lead to frayed nerves after the first few months of living with dust,

The owners of this loft bought the existing fixtures and modified it to suit their purposes. They constructed a wall behind the desk, where none had been previously, to create a bedroom.

rubble, and mayhem. A lot depends on you—your temperament, life style, work habits, and so forth.

Even if you don't have cash in the bank, you may be able to get a home-improvement loan if you have a good credit rating. This can allow you to do some basic work at once so you can move in, or to pay a fixture fee for a completed place.

SHOESTRING-BUDGET LOFTS, LUXURY LOFTS, AND POINTS IN BETWEEN

Converting raw space into finished or semifinished living areas can be an expensive proposition, ranging from a minimum of $1,000 to $20,000 and more. Nonetheless, loft renovation can be done in a simple way on a modest budget by the average do-it-yourselfer. The sheer size of most

This kitchen—simple, esthetic, and efficient—was inexpensive to do because built-ins were kept to a minimum, water and drainage supplies utilized existing layout, and all appliances were second-hand.

Appliances can be scavenged off the street. This refrigerator got a face-lift with some paint and a wooden door handle.

lofts requires investing in vast amounts of raw materials—sheetrock, wood studding, and lumber—which are very costly. Resurfacing walls and ceilings can take prodigious labor. A lot depends on your personal criteria of what is and is not acceptable for you to live in.

If you are on a limited budget, you should think in terms of doing as much as possible yourself. Keep your subdivisions and level changes to a minimum and use all existing facilities if feasible. For example, if the area has a functioning toilet and sink, use them and perhaps build around them to make them more attractive. Likewise, keep your kitchen area close to existing water and drainage supplies. In addition, you can buy used or scavenged stoves, refrigerators, tubs, and so forth. Any used goods have their drawbacks, such as outdated electrical systems that use a great deal of electricity or that may break down and be impossible to repair. A high degree of finish in a short period of time is not usually feasible on a low budget. You may prefer sanded floors that have been restained and polyurethaned, but painted floors will give your loft a finished look and not cost you as much.

If you have a large amount of money to put into renovating a loft, almost anything is possible. With the large spaces involved, fabulous visual effects can be achieved with imagination and money. It is still possible to strike a median between the two extremes and have a beautiful loft.

How much money you will need to spend on renovation depends on your personal skills as well as the amount of free time available to you. It is always cheapest, if you have any competency in a particular area, to do the work yourself. Hiring others will immediately increase the cost. Nowadays 60 cents of every dollar spent in construction projects goes for labor and 40 cents for materials. Certain complex tasks should be done by knowledgeable professionals. Having the job done right the first time is well worth it and can save you problems, such as haggling with inspectors over such basics as main breaker boxes for electrical service and heavy plumbing work on main drains.

There is also the factor of time. If you are working five days a week and are planning to work only on weekends and evenings, it will take you a long time to do major renovation. If you can afford it, you may be happier calling in a crew to do a lot of the work at once. Living in the middle of a construction project is not really fun.

If you are planning to do some of the work yourself and to call in professional help on other tasks, be prepared for the inevitable gaps of time when nothing gets done because you are waiting for the electrician in order to go on to the next step or because materials have been delayed due to shortages or strikes. Professional plumbers and electricians—in fact, all contractors in general—tend to hop from job to job, and even if you are ready for them on Monday, they may not have time to do your installation until the end of the week.

Delays of all kinds are quite common in loft renovation. That special sink you want may be back-ordered eight weeks from the factory in Ohio. Or, as happened to a friend of ours, a toilet may be factory-defective and involve weeks of argument back and forth over who is to blame. Initially, our friend was able to get the exact toilet he wanted —an expensive one-piece unit—but after it was installed he discovered a tiny hairline crack in the casting of the vitreous-china bowl-tank assembly. The crack was so small that the leak at first appeared to be only moisture condensation, but the dampness became more serious after each flush. He called the store where he purchased the toilet. The owner immediately turned the problem over to the manufacturer, whose closest field office was 50 miles and a long-distance phone call away. The first few calls established that the problem could be handled only by the manufacturer's roving on-the-site inspector, who was responsible for a large metropolitan area and who hardly ever came into the office since he was on the road most of the time. Our friend was told to leave his name and number and he would be called back. There was no response at all for two weeks. Another call uncovered the fact that the field rep was actually on vacation but would certainly deal with the matter promptly upon his return. At this point our friend asked to speak to a supervisor and explained that he didn't feel like waiting any longer and that he wanted something done immediately, now, today! Mumbled apologies and another "We'll call you back." Suddenly, a few days later, he received a late-night call, long distance from the toilet manufacturer, asking what the problem was. It was explained once again, and our friend was informed that he would be contacted soon and told where to pick up the replacement. Some weeks later he was directed to the supply house 50 miles away—the only place within the metropolitan area with one of the deluxe units. Who had to spend a day traveling to exchange the toilet? Who paid for the labor involved in removing the defective unit, renting a vehicle to get to the supplier and back, and reinstalling the new toilet? The manufacturer? Forget it. Our friend did—not to mention the long-distance calls and the most valuable of all things, time. So, as with the financial projection for the renovation, you should figure your approximate time and double that too.

.4.
Space Planning

If you have never lived in a loft before, you should be prepared for some very real differences between apartments and lofts. The most basic difference is in the amount of space available to you and the openness of that space. This involves some rethinking about space, since most of us are used to living in small modules. Given a large amount of space, most people get a little crazy and plan a series of rooms for various activities. Although some functions demand separate partitioned areas, it is a shame to ruin a big space with a lot of subdividing. Undeniably, most people like the bathroom area to be private, separated from the rest of the living space, but few other activities need total separation. Many people seeking out loft space have several different interests or activities they want to pursue. Often they think they must have separate areas for each. If you are involved in carpentry or sculpture, it might be preferable to have an enclosed area isolated from the living section. You don't necessarily have to build a completely enclosed room just for sewing, writing, or meditation. If you have children, you may very well want to build a totally soundproof room to get away from it all. In fact, it would probably be better to build a bedroom for them as well as one for yourself, because even in a 100-foot-long loft there is little privacy for most adults from the inquisitive eyes of children.

Bear in mind that partitioning costs money, and the least amount you can get away with is generally preferable. Why kill the feeling of open space? Although many people can visualize what they want to do with a certain space, others have trouble figuring out how things will actually look. One way to help the visualization and planning of your loft is literally to camp out in the raw space and get a feeling of it. If there is an existing toilet and sink, you can bring in some sleeping gear and a hot plate and spend a few days and nights getting to know the loft. You may have seen the loft only at one time of the day. Seeing it under different circumstances—different lighting and possible noise conditions—can influence your choice of where to put your sleeping or sitting areas. If you like to wake up to the sunrise, you could put your bed next to a window that gets the first morning light. If you are a night person, you may want to locate your sleeping area in the darkest, cosiest corner of the loft, thus enabling you to sleep late, unaffected by the harshness of daylight or the clamor of trucks on the streets below. This sort of advance planning is especially useful if you have decided to hire a crew to follow your plans rather than do all the work yourself. If you are not going to be there during the construction, it is very important to have a clear idea of what is going to take place during your absence. Coming in at the end of the week and discovering that the closet is not quite where you wanted it can be very costly (if you decide to change what has been done) and can involve aggravation all around. Apart from the extra money it takes to destroy and rebuild, contractors hate to backtrack and do things over again.

We know a couple who had some great ideas for their loft. However, they had not spent much time in it before the renovating began. On paper it looked like everything they wanted would work out perfectly, but once construction was completed some of their ideas fell short of their dream or just didn't work. They then had to either rethink what was there or tear down and start again. They chose the latter, and it wound up costing much more than they had thought.

If you are going to do some partitioning, and some may certainly be in order, think carefully about doors. Do you really need them? Doors are good for areas of total privacy and possibly for storage space. But areas can be made somewhat private without doors, simply by constructing carefully placed walls. If your ceilings are reasonably high, you can make the walls free-standing—that is, stopping at 8 feet or so rather than going all the way to the ceiling. Free-standing walls allow for the passage of air and light and give the illusion of still-open space, while at the same time providing some separation for one area or activity. Some people, particularly writers, need a sealed-off area that is isolated and soundproofed from all outside and household noises. This is a matter of personal preference, as is the whole loft-living concept; everyone will suit himself on these things.

While you're planning the layout of major areas such as bathrooms, bedrooms, and kitchen, don't forget to leave space for closets and storage. If your loft is small, you should think about utilizing the space around stairways and pipes or under loft beds—space that might otherwise be unusable.

Another major difference between apartments and lofts is in the kitchen and bathroom space. In apartments these facilities are usually designed to take up the least amount of room in the most efficient manner for the "majority" of people—but this does not necessarily mean it is the most efficient for your needs. In a loft you can set aside as much space for kitchen and bathroom (or rooms) as you are willing to give up for other functions.

Areas for different activities don't have to be completely enclosed. This writer's alcove with its free-standing wall allows physical isolation without cutting off air and light.

Because of the limited space in our present loft, our kitchen and bath are rather small. In our next place we plan to devote a lot of space to the kitchen, since we're both into cooking and since the kitchen seems to be the place most of our guests end up.

Since lofts are often not centrally located, or are located in commercial areas, chores that were easily accomplished in your apartment or house now may become more difficult. This is especially true for getting laundry done. Unless you install a washer and dryer in your loft, you might have to hike many blocks to get to a laundromat. You might want to plan an area (not necessarily a room) for these items, especially if more than two of you will be living there.

Garbage disposal is another chore that becomes more difficult once you move into a loft. Commercial areas in many cities do not have municipal collection; therefore tenants have to make arrangements with private garbage collection companies for refuse removal. If you have fallen in love with a fifth-floor walkup loft, you may want to leave room in your kitchen for a garbage compactor to reduce your volume of garbage—although you will still have the problem of disposing of a heavy, compacted block of matter in a plastic bag.

In planning the layout of your loft, try to work with the placement of existing elements, such as plumbing sources. It is far less expensive and time-consuming to put your toilet, shower, and kitchen close to the source of existing plumbing than to put them on the opposite side of the loft—which would necessitate running many yards of expensive pipe and raising a platform to drain the area and to conceal the pipes. If money and time are of minor importance and the idea of a raised kitchen island excites you, then by all means put it wherever you want.

If you are on a budget and are going to put bathroom and kitchen facilities near the source of the old plumbing, keep in mind that you do not have to have access to your bathroom through the kitchen. (As mentioned before, you can utilize an old toilet flange for a shower drain, as long as the old drain is not blocked. The mechanics of shower installation are discussed on pp. 67–75.) The alternative is to put kitchen and bathroom back to back. Also, plumbing noises are then restricted to one area of the loft.

Other existing elements to take advantage of include shaftways and window locations for exhaust venting in kitchen and bathroom. If possible, locate your kitchen next to a window or at least in a place where you can vent directly through a wall or through the roof.

Columns are something that you never come across in apartment buildings, but they are often found in lofts. These are structural and cannot be removed. Therefore you will have to work around them, or work them into your layout. Interesting effects can be obtained by painting them in bright colors; or you can paint them white to make them as unobtrusive as possible. You can also utilize the columns by hanging hammocks between them. Whatever you choose to do, they are permanent fixtures and must be dealt with as such.

The layout of this loft (*right*) called for the kitchen and bathroom to be far away from the drainage and water supply stacks, so it was necessary to run long lengths of pipe (*left*) and construct a platform to cover over the plumbing and to compensate for proper drainage pitch.

Planning as far in advance as possible can help you save money in the future. If you want a washer and dryer, you can plan ahead for plumbing, electricity, and space and then install the units themselves at a later date. The same is true for air conditioning; since most lofts are larger than apartments, a large-sized commercial unit is usually needed to cool the space. You'll need a separate 220-volt line specifically for the unit. Here again, planning ahead has its obvious advantages —for example, you can run the line during construction and conceal or paint it over, rather than having to mess up a ceiling or wall at some later time when you've finished painting and decorating. The same applies to bathrooms and darkrooms; both require plumbing and electricity. If you want them to be expandable, try to plan ahead accordingly. Oversized and custom bathtubs need more than the normal amounts of hot water; if you are providing your own hot water, you'll definitely need more than the usual 30-to-40-gallon heater—a 50-to-75-gallon unit with fast recovery would be essential. Also, for huge tubs, you'll need oversized or multiple water-fill valves to allow for faster filling time.

An increasing number of people want steam machines, saunas, and Jacuzzis. All of these units need heavy 220-volt electrical supplies, and it is definitely preferable to install these hookups at the time of any initial electrical work. Steam machines can be installed later, as long as you have some undercounter cabinet space available, along with the 220-volt hookup. Saunas tend to be in separate enclosures and can be added later, as long as there is a plumbing and electrical setup ready for them. Jacuzzis are installed during initial construction.

This tub was custom built and is equipped with steam machine, double water-fill valves and handset for faster filling, and a pair of high-recovery rate water heaters.

Along with the feeling of linear space—width and length—in a loft, you have the additional feeling of height. In a normal apartment house you certainly don't get to think much in terms of ceiling height (apart from the fact that it's probably too low). In a loft it's often a whole new world. Most lofts have much higher ceilings than the average residential apartment building. Heights from 12 to 20 feet are not uncommon. This brings into play the notion of potentially different levels. Some lofts with very high ceilings can actually look a bit out of scale, especially if they are long and narrow, as many old loft buildings are. Creating different levels within the space is one way of alleviating this feeling, while at the same time utilizing the vertical space in an interesting manner. Even if you don't need the additional living and working space (which may be necessary in a small place that has high ceilings), it's an exciting way of treating space. (The mechanics of changing levels are dealt with in detail in Chapter 12.) In our ordinary day-to-day experience we see little if any multilevel activity, except in houses. In a loft whole new floors can be created if desired. Even a modest level change of a foot or two can make a difference, and it can be accomplished much more easily than the addition of a floor. The raising of a bathroom floor may in fact be necessary if your existing drainage points are not low enough to place a tub or shower directly on the floor. As long as you are doing this, you can raise the floor a little higher and create a sunken-tub effect.

This brings us to the question of fabricating bathroom units yourself versus buying manufactured units. It used to be that to build your own bathtub you had to lay down a heavy cement foundation with a lead underpan and then lay in conventional mortar mix and tiles. This can now be avoided, thanks to modern boat-building materials—epoxies and fiberglass resins. You may or may not want to do the built-in work yourself. We wouldn't recommend it if you are not adept at carpentry and basic plumbing at the very least. Built-in units of any kind, be they kitchen, bathroom, or living room, can be terrific, but unless you are going to do all or most of the work yourself, they can be very costly. They can also take a lot of time, because all irregularities of the structure have to be carefully worked out. Ready-made, standard pieces have definite advantages—including price, availability, and ease of installation. If you are willing to accept stock-size kitchen cabinets, you can probably obtain them easily and install them quickly. The same is true for bathroom fixtures; standard units are readily available and are much less expensive than high-style exotic pieces. Again, this is a matter of individual taste. If you want exotic stuff, be prepared to wait six to eight weeks—or even longer for such things as imported tiles and rare woods.

Building in looks very good, but remember you can't easily take it with you. If your lease is shaky or nonexistent, you may not want to bolt down too much in case you have to pack up your investment on

A handy loft dweller built this couch for listening to music and watching TV.

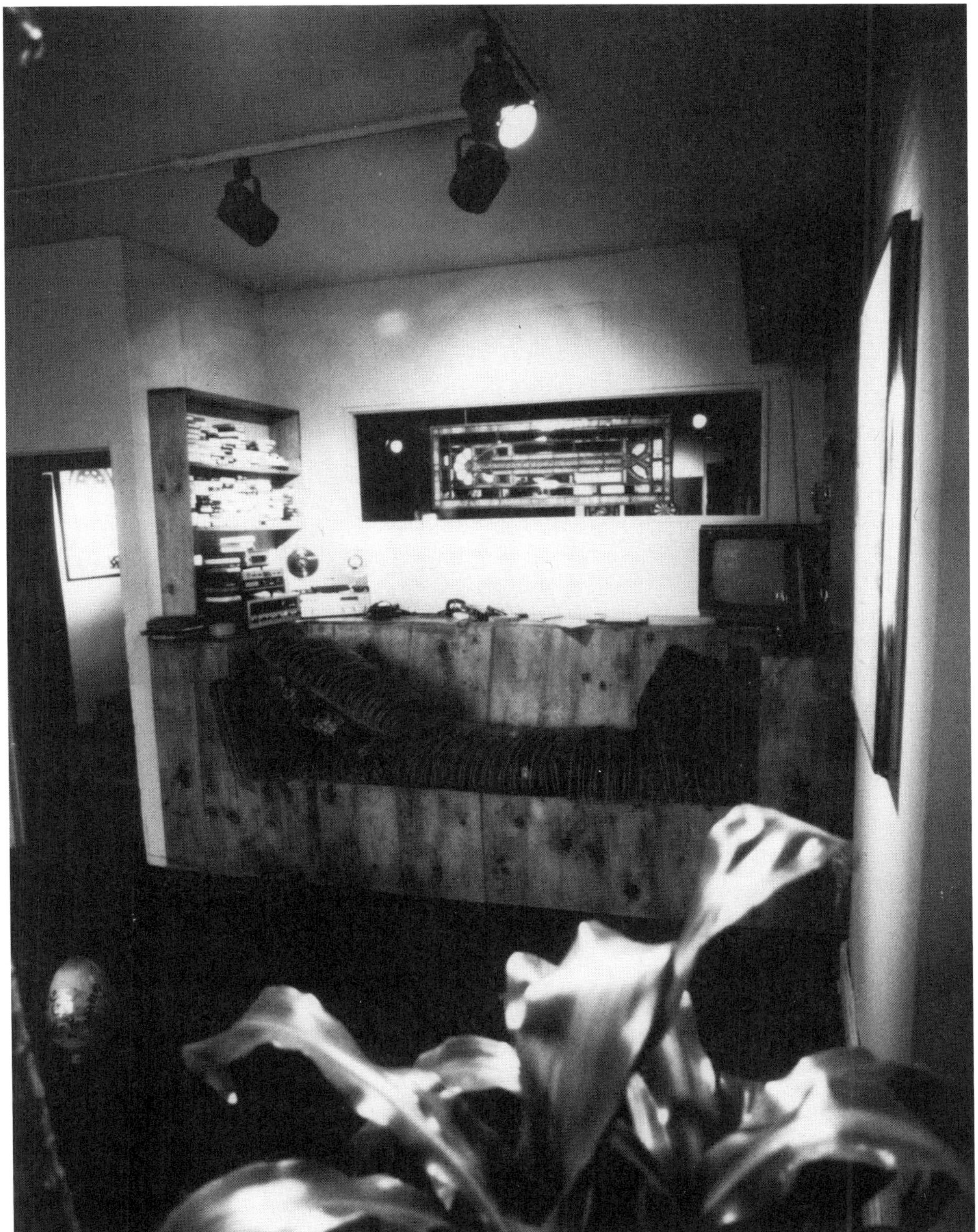

short notice. You try to avoid this situation, of course; but eviction can happen, and it may give you some pause. If your lease is secure enough and time and money allow, then going the built-in, handmade way can give an otherwise unprepossessing loft the finishing touch. Also, because the fixture-fee system works in many cities, you can pass on the costs of customized built-ins to the next tenant and use the money to move into another place and do it all over again!

.5.
Doing the Work Yourself vs. Getting Professional Help

Undeniably, one of the great satisfactions of living in a loft is the fact that you can build an environment tailored exactly to your personal needs. Many people derive great ego satisfaction from actually doing the physical work themselves and carrying it through to completion, just the way they had imagined it or put it down on paper. One of the harsh facts of life, however, is that not all of us have the ability to simply put it all together. Some people find it impossible even to drive a nail straight; others, given plenty of study, training, and on-the-job experience, can build houses from scratch. For many the path lies in between. Many people in today's industrial-mechanical society have reasonable levels of skill with various tools and materials.

Whether or not you will be able to renovate a floor by yourself depends on your level of skills, the amount of time available to you, and the variety and type of tools you possess or have access to. Even for the most skilled and experienced, there are certain heavy jobs that will require professional help—such as cutting into cast-iron main drains to provide new drainage points. The equipment and tools needed for this job are very expensive, and unless you plan to use them all the time, they are impractical to own. Since most people have this job done only once at the outset, it simply does not pay to buy the gear necessary for the job. Mechanically speaking, anyone could learn how to do the work—it does not require exceptional dexterity. However, there are numerous little tricks of the trade to learn, and unless you work as a plumber's helper for six months or so, forget about doing some of the more complex plumbing tasks. There are plenty of other plumbing jobs—such as sink, tub, and toilet installations—that almost any loft dweller can handle with a minimal investment in equipment.

Another area best left to a qualified professional is basic work on electrical installations. Once you have service on your floor, you can, with a reasonable amount of manual dexterity and a basic understanding of electrical principles, run most of your own wiring via such materials as Romex, BX cable, or rigid metal tubing (depending on local electrical codes). Most codes specify that work be performed by licensed electricians, but you'll be OK generally if you simply comply with the code-specified materials and do the job right—that is, using the right-sized wires and fittings and not trying to skimp by overloading the circuits and creating a potential fire hazard.

As far as doing partitioning, flooring, and built-in work, a lot depends on your skills and the time available to you for working on your loft. Some people are really turned on by the idea of doing everything themselves. That's fine—getting directly involved in the project is part of the loft mystique. Other people are frightened by their own lack of skills but are nonetheless eager to live in a loft. There is no shame in getting help for any and all aspects of loft renovation. For some, it may be the only way they can do it. In fact, an ever-increasing number of loft dwellers are people who hold nine-to-five jobs—people who are not necessarily artists but are simply looking for an alternative to the restrictions of apartment living. If you are one of these people, you may want to call in a whole crew and get the job done in the shortest possible time. This route is also good if you really don't want to bother with all the little details and problems of putting a loft into shape—running around for supplies and materials, coping with dust, confusion, chaos, and so forth.

It will cost more to go the route of contracting out the work. If you know enough about what has to be done, you can be your own contractor; that is, you can hire all the different tradespeople on your own and show them what has to be done. If you're not quite sure where all the things should go, you may want to hire a contractor to take care of hiring and supervising general workers, specialty craftsmen, and licensed professionals and making sure the construction takes place in the correct sequence. (This important aspect is detailed in Part II.) If you are not knowledgeable on construction procedures, you are going to be at the mercy of a contractor. Hopefully you will be dealing with an honest and reputable person, but you can be taken rather easily by the few disreputable types in the business. Be careful when working with a contractor; word-of-mouth referral is one of the best ways of ensuring that you will not be ripped off. Also, if you can, bone up on the terminology used by these people. Be specific where you can; use correct nomenclature—it impresses the contractor if you can use his language, and he'll be less likely to rip you off if he thinks you understand what he does.

Another area in which to use prudence is financial arrangements. Resist hard-luck stories and stick to the standard contracting method of payment: 33 percent up front, another 33 percent halfway through the job, and the rest upon completion *to your satisfaction*. Withholding the final payment is the only really practical way of ensuring that the job gets finished. If you are contemplating a big job that involves a good deal of money, it may pay you to put the job out for competitive bids. Bear in mind, as is true in all bidding situations, that the lowest bid does not necessarily mean the best job. If you get at least three bids, you will have a good idea of what to expect.

Along with the rapid growth in loft living, there has developed a whole new type of contractor, one who deals specifically with lofts and their unique problems. These people are very often loft dwellers themselves and are thus very familiar with the ins and outs of the buildings. Also, they are generally much cheaper than regular contractors who do "standard" construction jobs at grossly inflated union wages.

The free-lance writer or photographer who takes on small construction jobs to supplement his income is not unfamiliar in most loft areas. Hiring such a person to work on your loft does indeed have its advantages—namely, familiarity with the type of work necessary and relatively low cost as compared with other workers. In fact, depending on the type of work you are contemplating, area people may be the only type of help available to you. Another factor in favor of hiring local people is that you are helping the loft subculture by recirculating money among loft dwellers. Local workers can act in a contractor's capacity for you if you wish; or they can work on a day-to-day basis as you need them. If you are a handy type and are doing most of the work yourself, you still may need an additional body from time to time to hold up the other end of a pipe or a piece of sheetrock. If you can't get your friends in on this, you may be able to get some day helpers from a local college or university. Students are often very energetic workers, having been cooped up in classrooms day after day, and they are often happy to do something physical and to get paid for it as well. Many students are short of cash and appreciate the money.

The son of a friend of ours is an architectural student who volunteered his services on one of our projects as a means of gaining first-hand, practical experience. He had spent almost five years reading about architecture and construction but learned some invaluable lessons by doing part of the actual construction himself. He certainly learned how physically demanding the work can be and gained respect for the people who do it.

If you are not 100 percent proficient in executing the work you have outlined for your loft, keep in mind that you may not actually save money by doing a job at which you are not skilled. Sometimes attempting to do the plumbing yourself can cost you twice as much as calling in a professional from the beginning. First, you have spent money for materials that may end up having to be ripped out and discarded. Second, you have to incur the cost of new materials plus the plumber's fee for doing the job over and ripping out the work you've done incorrectly. So you should give a great deal of consideration to the possible savings involved and make your decision accordingly. You may find out that the difference is not very great on some jobs and that you'd be better off getting a professional. The decision requires accurately assessing your level of skill and competence. It's tough to have to admit that you are not the master builder that you envisioned yourself to be, creating and executing every single piece on your own. Perhaps you should be happy with being able to communicate your thoughts and fantasies about your loft to someone else, rather than trying to do it and have it fall short of your dream.

.6.
Tools and Materials

TOOLS

Assuming that you are going to do at least some of the work yourself (such as partitioning, platforms, some plumbing and electrical work), you're going to need quite a few tools. Remember that few loft buildings have a superintendent to call up every time some item goes on the fritz, so it's a good idea to have at least some basic tools on hand for those little midnight repair jobs.

If you've got some skills already, you probably own a few basic tools, such as saws, hammers, and screwdrivers, although you never seem to have enough of everything, no matter how many tools you accumulate. The deeper you get into construction, the more and more specialized each tool becomes: a circular saw is augmented by a saber saw, then a router, then a power planer, and on and on. We've discovered that one can go too far, but a certain minimum amount of equipment really makes the work go better, faster, and easier.

For example, if you are going to be doing a fair amount of framing and studding in wood, you'll benefit from buying a 20-ounce hammer instead of the standard 16-ounce unit. You'll be able to drive nails a lot longer and with less effort. In general it pays to spend a little more money on high-quality tools, especially if you are planning a more protracted or elaborate job. It is a tremendous waste of money to start out with a cheap power saw, have it burn out on you, and then have to buy a heavier unit. The same holds true for drills and other equipment. Most power tool manufacturers make at least three grades of tools. For example, the cheapest grade of saw is for "hobby" use and is fine for occasional jobs—the type of work the average suburbanite might do on weekends cutting shelving or building cabinets for a playroom. A saw of this quality is usually in the $10-to-$20 range—and that's about how much use you'll get out of it if you try to build your loft with one.

Then there's a sort of middle-quality saw, which runs anywhere from $30 to $60; it is quite adequate for serious building and won't burn out the first time you hit a tough knot in the wood. Given reasonable care and luck, this tool is sufficient for the average loft renovator who is trying to get his place together.

Top-quality professional or tractor's grade saws are usually quite expensive—anywhere from $80 to $150—but are built to last almost forever. They are capable of day-long use without overheating and are essential for jobs where speed and first-class finish are paramount and cost is secondary. You probably will not need this top-quality tool, unless you plan on doing a lot more work later on, or perhaps building a house from scratch in the countryside—or unless you happen to be a tool freak and just appreciate the smooth and efficient way these top-of-the-line pieces of equipment do their job.

Also, bear in mind that almost any kind of heavy-duty tool can be rented for a few days for small jobs. A big drill for drilling into concrete sometimes costs over $600; it would be foolish to purchase one unless you were going to break up concrete for a living!

If you're going to do some of the work, you'll need the following as a minimum:

1. *Plumbing.* For basic jobs you'll need a big pipe wrench (good for up to 2″ pipe), a couple of smaller wrenches, water-pump pliers (Channelocks), hacksaw, and basin wrench. If you want to do work in copper tubing, you'll also need a propane torch, striker, solder, flux, and tubing cutter.
2. *Carpentry.* Invest in an assortment of hammers, screwdrivers, and pliers, several heavy-duty grounded electrical extension cords, a hand saw, electrical circular saw, saber saw, drill, carpenter's square, ruler, plumb bob, line levels, chalk lines, and a set of chisels.
3. *Electrical work.* If you're planning to run thin-wall metal conduit for your wiring, you'll need a pipe bender for each size tubing used (the most common are ½″ and ¾″), a good pair of lineman's pliers with insulated handles, and a circuit-testing device. If your code allows the use of Romex (plastic-clad) or BX (metal-clad) cable, you'll need a sharp hacksaw to cut the stuff, wire strippers, a pair of pliers to strip the wire and bend it, and a skinny screwdriver to connect whatever device you are installing.

A word of advice on using an extension cord with power tools. Make sure the cord is heavy enough to accommodate your heaviest tool. Use a three-wire extension cord, with a three-prong plug at one end and a three-hole socket at the other. If you use a two-wire extension, the tool will not be properly grounded and you will not have protection from electrical shocks, unless your tools are "double-insulated," thus obviating the need for a ground wire.

MATERIALS

A few words on the acquisition and assemblage of materials used in the average loft. If possible, try to find a few reliable sources for the different materials necessary and keep using them throughout your job. For example, try to strike up a good relationship with your friendly neighborhood lumberyard. You may even be able to get a break on prices if you can convince the owner that you are going to be spending a certain number of dollars at his yard over the next six months. It's worth a try, and even if you don't get the break, you'll get used to the way they do things and vice versa. Employees might begin to recognize you and give you some helpful information if they see you are a serious person who is truly interested in what they have to tell you. You can learn a lot from people who work in lumberyards; many of them have been working with wood for decades and know a lot about it.

The same applies to your plumbing and electrical supplies—try to find a wholesale house in each of these specialties that will sell retail to you at a contractor's discount, which can vary anywhere from 10 to 40 percent off. Again, shop around at various places. Sometimes the most convenient place may be both overpriced and have surly, unattentive workers serving you. Also, some big places don't want to know about the little guy with the set of problems that lofts present. On the other hand, you can often find a place with employees who will offer you a perfect solution to your problem and who don't mind selling you the part either.

So look around for supplies and get the best you can. Also, plan ahead carefully so you can buy your supplies in large quantity and take advantage of any wholesale discounts on the material. Apart from the potential savings in buying in bulk, you may find that particular materials, such as knotty pine tongue-and-groove planks or tiles, are in short supply or only available now and then. If you run out halfway through your job, you may be told it's a twelve-week wait. Or, as happens fairly often, you may find the item is discontinued forever and then you'll really be stuck. So if you're able, plan to get just that little bit extra; various suppliers have rules of thumb for estimating overage. For certain types of woodworking it'll be 20 to 30 percent or more; for tile work you may need more than that to compensate for potential breakage, cutting corner pieces, and so on. The same is true for electrical wire, Romex, and BX cable. Calculate on the generous side and don't forget to figure at least an extra foot of wire at each connection point—that is, each outlet or switch—to avoid running out.

If you have a truck or van at your disposal, you may be able to buy your materials at some saving by going to a large suburban supply depot. (Most urban supply houses are more expensive, if for no other reason than that they have a higher overhead in the form of rent, labor, trucking, and fuel, which they have to make up for in the pricing of the goods they sell.) Most suburban places are set up to take care of requirements for home remodeling and repair, but they carry a lot of the same materials that loft dwellers need—studding, sheetrock, electrical cables, plumbing parts, and so on. By going in person you can see what you are getting and thus run some sort of quality control over what they give you—especially in the case of lumber. Crooked 2″ × 4″s are more easily replaced on the spot in the lumberyard than on the street in front of your building, as happens when you phone in an order and get a sidewalk delivery. Having your own transportation also enables you to shop around for the best possible prices and service. This can take time; but if you have plenty of time and not too much money, it certainly makes sense to do so. Another way of saving on materials is to remove existing studs and electrical cable carefully when you do your demolition work. Some of what you take down can be reused in your renovation for one purpose or another.

One way of saving money is to buy used or second-hand appliances like refrigerators and stoves. Bear in mind that buying used water heaters can be a false economy; it's true they have few moving parts, but they do have delicate fiberglass dip tubes as well as thermostatic temperature controls which can be easily damaged if handled roughly and are quite costly to replace. Used stoves can be easily restored and, if properly cleaned, will work well for long periods of time. If you purchase used items from an appliance dealer, try to get a written guarantee on the bill of sale, say for at least 90 days. This will prevent you from getting stuck with a refrigerator that expires after the first few days. If you're buying from an individual through an ad in a local paper or supermarket bulletin board, then you'll simply have to take your chances. Common sense in these dealings will probably get you through. Many people have fabulous old 1950s fridges that still run like fine old clocks, whereas some unfortunate souls have modern units that require the bimonthly attention of a service technician who charges $17.50 to walk in the door and $15 per half-hour for labor. We even know people who have scavenged stuff right off the street that is still in good working order but too "old-fashioned" for someone else. Used sinks and bathtubs can often be found at the local dump and, if not cracked, are quite serviceable once you install new fittings and seals.

If you live in an expensive electric-rate area (the highest of which is New York City), you may wish to give consideration to the energy consumption aspects of new versus old appliances. The electricity consumed by an old, inefficient unit can be as much as double that of a new unit (this does not apply to frost-free refrigerators, which use about twice as much electricity as new nonfrost-free units). The recent energy crisis has created a new awareness of electrical consumption with regard to all appliances, not just refrigerators. By using higher-

Carefully removed existing materials can be reused in your renovation.

quality motors and compressors, it is possible to conserve and reduce electrical consumption.

In general, think very carefully when planning and buying materials for your loft. A little advance thinking can save you from the sudden realization, halfway into the project, that it's not quite what you have in mind and that you want to tear it all out and start over again. You're the one who is going to live there; you'll want it to be right the first time, since your labor and energy are involved. If you install a unit and then have to remove it and/or rebuild it, you are just increasing your total outlay for materials.

PART II
ORDER OF CONSTRUCTION AND TECHNIQUES

After lath and plaster have been torn away, frame is exposed and ready for stud removal.

.7.
Demolition

Before you do any construction, you should remove any unnecessary nonload-bearing walls. In fact, the time to do all demolition is in the beginning. It is messy, dirty, and destructive work, and newly built structures could get in the way and be damaged. Most nonload-bearing partitioning consists of wood studding and lath and plaster—or sheetrock and metal studding—that is easily removed. There is no quick way to get these walls down; your best bet is to start around a door-frame opening and take all the molding away with a wrecking bar. Once you've exposed the frame itself, you can get at the studding and start to knock each of the studs out from the sole plate. If you have old-fashioned plaster-and-lath walls, you'll have to crash into the wall with a heavy sledgehammer and break it all off the studs. Always start at the top—it's no fun to have a bunch of loose plaster fall on your head, even if you're wearing a plastic hard hat, which we highly recommend. Once you've stripped off one side of plaster, you can remove the rest of the beams by knocking the bottoms away from the plate, using a lever (see figure 1) or one of the beams to pry them loose. After all the studs have been removed, you can pry loose the sole and ceiling plates with a crowbar. Having an assistant to hold one end of the ceiling plate as you pry it loose makes this job easier and a little less hazardous.

Make sure you have disconnected any electrical lines that are embedded in the wall. If you are in doubt, shut off power to the entire floor. Be careful too not to cut any existing water or gas lines; apart from potential disaster, landlords don't appreciate it! Check to see that every line—plumbing, electrical, or gas—is dead before you cut through it. This involves a floor-by-floor check to make sure the line is not in use by some other floor.

Once you've demolished all unwanted partitions, you'll find yourself with an incredible amount of debris. Getting rid of all this can be a problem. There are several approaches, each with its own merits. If the

Figure 1.

An old beam can be used as a lever to bust out old studs. Insert the lever in front of the stud to be knocked out and behind adjoining stud and push away from you, exerting force. Once the stud comes loose from the sole plate, swing the stud back and forth until it comes away from the ceiling plate.

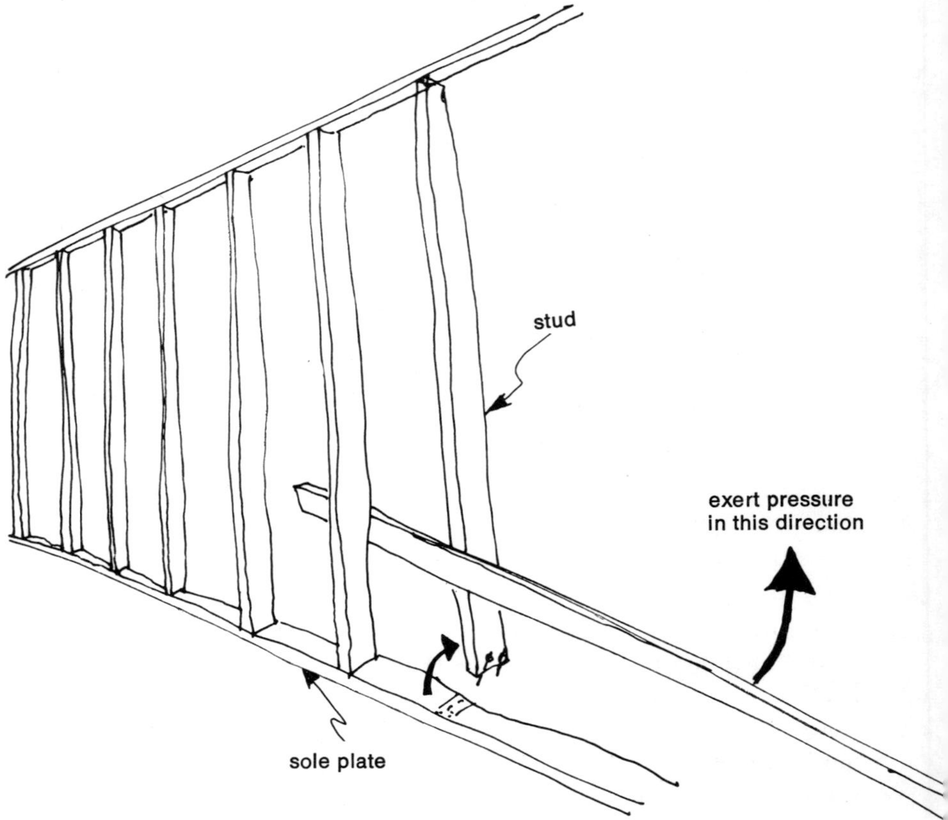

amount of debris is small, you can use the old method of sneaking a bag or box per day into your regular garbage pickup. (As mentioned earlier, in many commercial areas tenants have to make their own arrangements for garbage removal.) If there's a fair amount to be removed, you may have to truck it away yourself to the local sanitation fill or dump. Finally, you can pay a carting company to do it. Some carting companies will even do the demolition too. If you've never done demolition work before, you may find it exhilarating to smash down walls; or you may just prefer to have this particularly heavy job done for you by others.

STRIPPING AND SANDBLASTING BRICK WALLS

While you are demolishing partitions, you should also do any stripping off of plaster from brick walls, since you want to get all the plaster debris out at one time. Stripping walls to their original brickwork is common practice in the renovation of old brownstones, but it can be equally applied to many old loft buildings, which quite often have excellent brickwork. Brickwork, even if not in perfect condition, lends a very warm feeling to a place. If the concealing plaster is very weak and old, it can be removed easily with a hammer and chisel. If the plaster is stuck on well, it will be a slow job. A wide chisel and a five-pound sledgehammer will make the work go a bit more quickly, and a rolling scaffold assembly (which can be rented) can be very helpful if you are planning to do a large area of the loft. Heavy work gloves, a hard hat, and eye protection are useful for this type of work. Many people also use a dust mask for filtered breathing during demolition and plaster removal.

Once you've got the bulk of the plaster off, you will have to deal with a lot of dust and clinging bits of plaster. Go back over the entire area with a heavy wire brush and get as much as possible off by hand. The final treatment can be done with muriatic acid, which is applied with a large brush. Since the acid is extremely caustic, heavy rubber gloves *must* be worn as well as eye goggles and a thick shirt. With reasonable care you can avoid splashing yourself while applying the muriatic acid to the brick. The acid creates a chemical reaction (it fizzes when first applied) with the mortar and brick and "sets" everything a bit, thus reducing the amount of dust. If all this is too much, you can get a commercial service to come in and sandblast the exposed brick. Sandblasting will give the brick a far superior finish than the brush-and-acid method, but it is more complex and requires professionals with lots of machinery and huge pumps—and it costs a fair amount of money. A figure of $1,000 for a typical loft is not uncommon; however, sandblasting does look very nice and clean and definitely eliminates the falling-dust problem that is sometimes encountered with brick walls.

REMOVING CEILINGS AND EXPOSING OLD BEAMS

If you think there might be some old beams in the ceiling you'd like to expose, you can check it out by opening up a 3′ × 3′ area of the ceiling and seeing what's in there. If the beams look like they might have potential, you can pull down the old ceiling while doing your other demolition work. Again, a rolling scaffold can be a tremendous help in this situation. In addition to working in a more comfortable position, you'll be able to do the job faster and easier. If you are going to be working for some months, it may even pay to buy a scaffold setup; even if you're not going to pull down a whole ceiling, you may want to put up lighting tracks or do some extensive patching and plastering. A scaffold is also very useful while removing extraneous pipes or wiring conduit from the ceiling before you do your finishing and painting.

Once you've got the old ceiling down, you may want to refinish the old beams if they are particularly rough or blackened with age and dust. This is an arduous task, but a rewarding one. The beams can have a very nice luster by the time you are finished. Refinishing involves either sandblasting the beams or sanding them down with a heavy-duty auto-body sander and coarse paper, followed by passes with medium

A scaffolding is helpful to open up a ceiling to check for old beams that might be exposed and refinished. After finding beams unsuited to refinishing, the ceiling was closed up and patched.

Beams successfully exposed, sanded, and refinished.

and fine paper. If you want, you can coat the raw wood with spar varnish or polyurethane, or simply leave it. Naturally, you'll want to do all this work before you begin construction, since the area will be covered with tons of sawdust if you do sanding or sandblasting and certainly couldn't be occupied during the work.

Unless you plan to leave the ceiling exposed—that is, with the actual planks of the upper floor (or roof) showing in between the beams—you will want to install sheetrock or Homosote with sheets of fiberglass insulation between them. A completely exposed ceiling can look very nice, especially with sanding and refinishing, but unless you are on the top floor there's not going to be too much soundproofing or heating insulation between one floor and another. Both of these problems can be solved by installing thick (6″) glass wool insulation and then sheetrocking between the beams; this work will require a scaffold setup and plenty of time, but it is certainly worth it in multifloor dwellings.

If you are going to leave your beams exposed, you should consider decorating the room with sound-absorbent materials—heavy draperies, thick carpeting with underpadding, and the like. You might also call in an acoustical consultant to give you specific suggestions on soundproofing the area. This will cost a bit, but if you need peace and quiet to work, it might be money well spent.

Another way to cut down noise from one floor to another is to install a raised, false floor over the original one. Start by laying down a network of 2″ × 4″s on edge and filling the spaces with sheets of fiberglass insulation; then lay new wood flooring (pine, oak, or maple) over this. Sealing up all possible cracks and separations between floors and walls and ceiling can help a lot too. An efficient but costly way of soundproofing an individual room is to construct double walls (see p. 106).

Depending on the source of your noise problem, double- or even triple-glazed windows can help to cut down noise. Even simple tightening and caulking of loose window frames can be beneficial. Noise is often difficult to control because it can come not only from outside but also internally through either the air or the building structure.

There are various types of acoustical tiles available from lumberyards and supply houses for covering a ceiling as well as walls, but the esthetics of these may not be acceptable to all people.

According to engineers and contractors familiar with noise-control problems, it is usually better to soundproof the *source* of your noise with sound-resistant boards and carpeting than to install acoustical tile on your ceiling. For severe cases there are several different types of "white noise" machines available that effectively mask certain types of noises with less offensive low-level sounds. Experts contend that it is not actual decibel levels but rather the "annoyance factor" that bothers people. Your individual temperament will probably be the biggest factor in deciding what you can tolerate and where you will draw the line as to acceptability of noise levels.

Some friends of ours live on the fifth floor of a ten-story building. The ninth floor is occupied by a fellow who belongs to a rock group that practices late into the night, driving his fifth-floor neighbors, who seem to get the worst of the noise, crazy. The tenants on the eighth and tenth floors are not bothered as much by the noise. The way the building was designed, the noise seems to travel directly down to the fifth floor. Our friends ended up constructing and soundproofing a bedroom with thick foam-backed carpeting, putting lots of heavy fabrics down, sealing up certain windows, and putting special frames in the remaining ones. Some noise does still filter down to them, but they are able to live with it. They probably would have been better off, however, paying to soundproof the room where the rock group practices, thereby trapping the noise on the ninth floor.

EXPOSING OLD FIREPLACES

Many old loft buildings were once heated by coal- or wood-burning fireplaces, but as time passed the buildings were converted to steam and gas heating, and the fireplaces were simply bricked in and forgotten. Quite often during the stripping of a wall down to raw brick, an old fireplace will be uncovered. It is possible to resurrect such a fireplace, but some care should be taken in doing so. Amazingly enough, a significant number of people die each year from fireplace-related accidents involving sudden conflagrations. Old flues must be checked out for good draw—unobstructed passage of the smoke to a chimney that is in a good state of repair. Very often an old chimney is capped off at the roof; the cap must be removed or cut open to receive an additional sheet-metal device—a length of large-diameter (8″ to 10″) flue pipe with a rotator device to give better draw. Brickwork should be inspected for decay or damage and repaired. If you are not on the top floor of your building, you'll have to check out all the floors above to make sure you won't be sending clouds of smoke into the upper spaces. Once all these elementary but necessary precautions have been taken and your fireplace is functional, you can have something very special in the way of a genuine-antique operating fireplace.

If you don't have an existing fireplace on your floor, you can install an antique or modern free-standing stove or fireplace. You can vent directly through a wall or, if you are on the top floor, straight through the roof. There is quite a profusion of free-standing units available, ranging from simple old oil drums to lavishly decorated antiques selling for thousands of dollars—and you can take them with you.

Repairing old fireplaces can range from the simple repointing of decayed brickwork to the complete relining of a flue with new fire brick or ceramic flue sections. Repointing old brickwork means putting in fresh cement between the bricks where old mortar has fallen out. To do this, you will need some ready-mix cement, a hammer and chisel, and a couple of small trowels to apply the new cement. First you must chip out any loose pieces of cement with your hammer and chisel down to a depth of about half an inch. Be careful not to drive the chisel in too deep or you may jam it in and crack a brick. Work slowly and wear eye protection to guard against flying chips and debris. Once you've removed the loose mortar, you can mix up new cement and apply it with a small, pointed trowel. Or you can use one of the special narrow trowels sold for this purpose by masonry supply dealers and hardware stores. Wet down the area to be worked on and force the cement in with the point of your trowel. Fill the vertical joints first, then the horizontal ones. A mason's hawk (a flat piece of square metal with a handle underneath) is useful for carrying cement to the site and for catching any cement that falls away as you are applying it.

You will not be able to get inside the flue itself if it is in poor shape and needs work. In this case you may be better off inserting a piece of round or rectangular flue material from the roof above to act, in effect, as a new chimney. The new flue can serve the purpose adequately, as well as improving the safety of your fireplace operation.

.8.
Roughing in Plumbing

As soon as your demolition rubble is removed, you should start to do the rough-in or basic preparatory work to your plumbing system. Obviously, if you are going into a raw space you'll have to do more than if you buy a finished place and move in the next day—although in the case of the latter you may wind up redoing it all anyway. We will start with the basic raw-space situation and work up from there.

INSTALLING A SHOWER OR TUB

The first thing most people want to do when they move in is to install some sort of shower or bathtub setup. This is assuming you have at least a working toilet and sink already in. If you are willing to settle for a simple shower stall and are lucky enough to go into an ex-factory setup with two toilets, you can run your drain directly into the toilet flange of one of the existing toilets. You'll have to remove the old toilet, but this is readily accomplished. First, shut off the cold-water supply to the bowl. If there is a valve directly under the tank, shut off the water by turning the valve to the right until it's tight. If you don't have a supply valve, you will have to shut off the entire water supply to your floor. Next, remove the two flange nuts at the base of the unit and pry the whole thing up. The toilet assembly can be saved and used elsewhere, if you want, so be careful not to crack the porcelain as you pry it up from the floor; vitreous china, although strong, can crack under the wrong kinds of stress. If you are going to throw it out, you need not worry about breakage. In fact, the old flange nuts may be so corroded that you will have to chisel or saw them off; in extreme cases you may even have to destroy the bowl to get it up. Once the old bowl is out of the way, you will see a round brass flange and a hole about 4 inches in diameter. This is called a "toilet flange." Most showers and tubs can drain easily enough through 1½″ or 2″ pipe, so all you need is

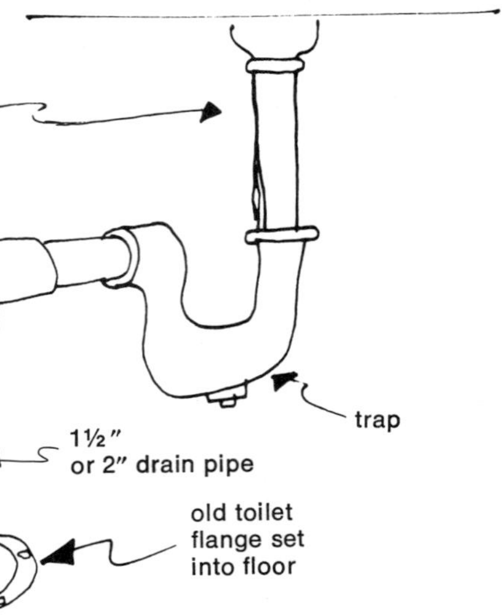

Figure 2.

An old toilet flange can be utilized to drain a tub or shower.

a reduction bushing from 4 inches to whatever size drain pipe you are going to use (see figure 2). The 2″ pipe might be preferable if you plan to put in a large tub or perhaps run a kitchen drain into it as well—all depending on your physical layout, of course.

A word on materials. You have a choice of several different materials in this type of work—cast iron, threaded galvanized iron, threaded brass, PVC, and sweat copper tubing. Check your local plumbing code first—it may require specific types of materials for certain aspects of the work. Plumbing today still utilizes old-fashioned, Industrial Revolution-age materials—namely, cast-iron drainage, made up of lengths of cast iron held together with tightly packed oakum and poured molten lead. Due to code restrictions, this method continues to be used in high-rise and institutional construction, despite the advent of threaded galvanized pipe and the new, easy-to-use PVC pipe and fittings. PVC is plastic pipe that can be used for drainage and cold-water supplies; it can literally be glued together and is simple to install. (A chemical reaction takes place as the pipe is cemented together with PVC cement, thereby creating a bond much like welding.) The old cast-iron method necessitated lots of heavy equipment for cutting the pipe, gas-fed portable stoves for melting down the lead, and so forth. All this can be done away with when you use PVC pipe. All you need are a tubing cutter or hand saw and a couple of paint brushes for the cement. The advantages to the loft dweller are quite apparent—PVC is a fast and permanent assembly, and low level of skill is sufficient to do the job. The usual range of fittings and adapters are available in PVC, and you can easily plug into an existing cast-iron or threaded galvanized system. Do not run water through the lines for at least 24 hours, as the cement takes a while to set properly. PVC pipe *cannot* be used for water over 150° F. Hot water requires a more expensive type of pipe, also available in plastic, called PVCP. There is one disadvantage to using PVC and PVCP: the pipe and fittings are not interchangeable, and care should be taken when working with the two not to mix them up.

On the other hand, sweat copper tubing can be used for hot- and cold-water supplies, as the same pipe and fittings are suitable for both. Sweat copper tubing and fittings, though expensive, are relatively easy to assemble—there's no threading to bother with (and if you've ever seen the machine needed to thread up conventional pipe, you know how tedious it is to make threads on each piece of pipe). Just cut the tubing to length, clean the ends with steel wool, apply a little solder flux paste, push the fittings on, and go on to the next fitting. When you've got a few fittings in place, you can solder them all at once with a torch and wire solder. For a small job, a few Bernzomatic tanks will suffice; if you are planning a really big job, invest in a portable acetylene tank along with a striker to ignite the torch.

A PVC toilet drain hookup.

If you can use PVC, it'll be easy work laying up as much drainage as you need; all the conventional fittings such as Ys, T-Ys, traps, 90-degree elbows, and 45-degree elbows can be used—you just make up what you need as you go along. If you are going to drain your shower or tub into a toilet-flange opening, place a short length of PVC pipe into the opening, pack in some steel wool, cement up the whole thing with some ready-mix concrete—and you're in. If you are not going to drain your shower into an old toilet flange, you may be able to run everything into an old sink-drain fitting; check the drain out well—if it's clogged, now is the time to have it reamed out thoroughly, before you start using it. Since old sink drains do not go into the drains at floor level, you'll have to raise your shower (or tub) off the floor as necessary for proper drainage pitch. Calculate a minimum rise of ¼″ per foot away from the point of drainage for proper pitch. The further away the shower is placed from the main drainage point, the higher off the ground it will have to be. If you have high ceilings and are planning a different level anyway for a bathroom or bedroom area, you won't have any problems. In any case you'll need a small platform for the shower or tub area.

Building a Shower or Bathtub Platform

Unless you are going to cut directly into a cast-iron waste drain, you will probably have to run your drain into either an existing sink drain or an old toilet-drain opening. For both of these situations you will probably find it easiest to put your new facilities on a slight platform. The illustration on page 71 shows a typical layout for running drainage from a shower or tub.

Since you're going to need some access under the platform, you'll need to build it at least 24 inches off the floor; when you've completed the hookups, you can fabricate some stairs to allow you to get up to the platform itself. Your construction can range from very simple to

A simple shower attached to a wall used existing drainage and didn't need any platform raising. Also note the PVC pipe and inexpensive utility valves.

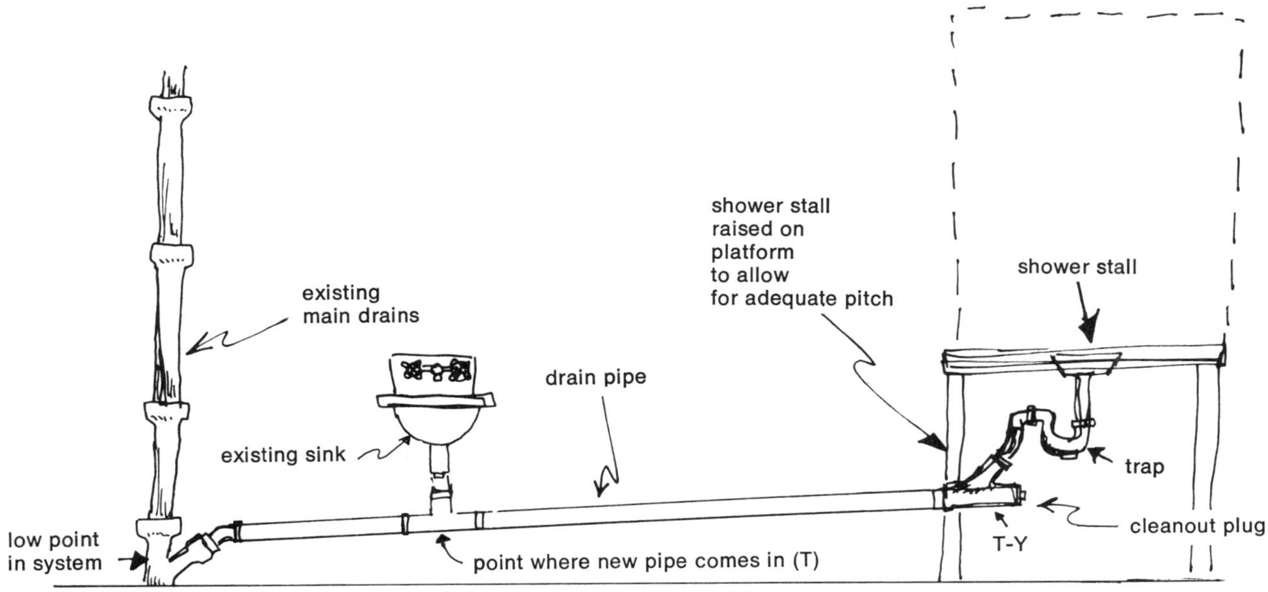

Figure 3.

Layout of shower being drained into existing sink drain. The drain pipe is connected to the sink drain by means of a T fitting.

very fancy, but the main thing is to provide a solid, level area to put the devices on.

The best way to support a shower stall is to fabricate a square platform consisting of two slabs of ¾″ plywood sandwiched over a frame of doubled-up 2″ × 4″s or single 4″ × 4″s. The two pieces of plywood should be glued together with waterproof glue, and the legs should be bolted to the base with carriage bolts and braced diagonally as well (see construction details in figure 4). The purpose of bolting down

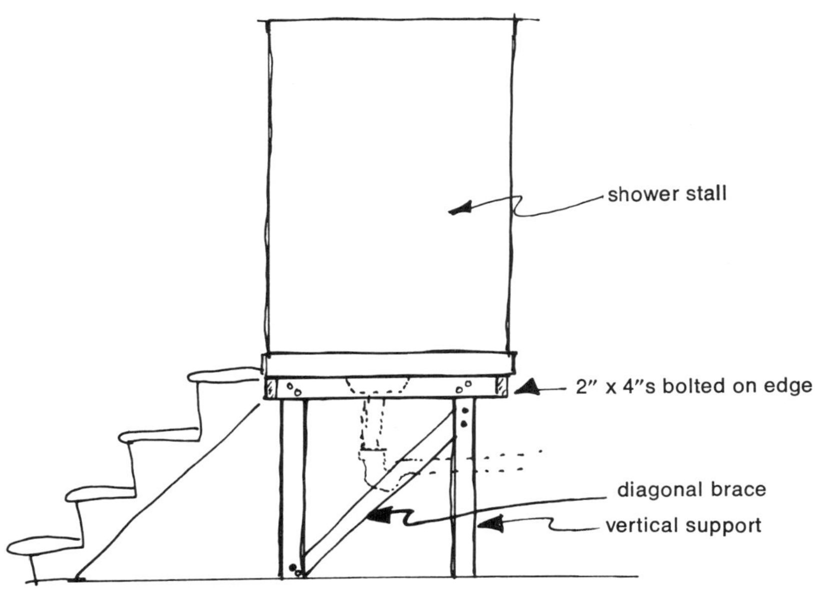

Figure 4.

A platform for a shower stall can be fabricated from 2″ x 4″s bolted together.

the platform is to prevent any possible movement that might loosen up the drainage connections and result in leaks.

Once you've got your platform constructed, you should set it in the shower space and attach it to the floor. Metal angle brackets and screws or short lag bolts are best, but you could probably get away with toenailing the legs to the floor (assuming you have a solid wood floor to nail into).

With the shower platform firmly attached to the floor, you are ready to assemble the shower stall itself. First, you should cut a hole in the center of the platform to allow for hooking up the base of the shower to its drainage point. Most shower-stall bases are made of concrete with a drain flange already set in; all you have to do is place the base over the platform, centering the protruding drain piece over the hole you cut in the wood. Since cast-concrete shower bases weigh several hundred pounds, you will need to call in some strong friends for this part. Once you've got the base in place, it will not need any attaching, for its own weight will make it immovable; don't worry about its slipping or shifting to one side.

Working now from underneath the platform, hook up a tailpiece from the shower base to a trap (which should be accessible in the future, as showers and baths will clog up eventually from the hair that invariably builds up) as shown in figure 5. If you are working in PVC drainage pipe, you should be using a PVC trap and fittings, as a galvanized or brass trap will put too much stress on a PVC system.

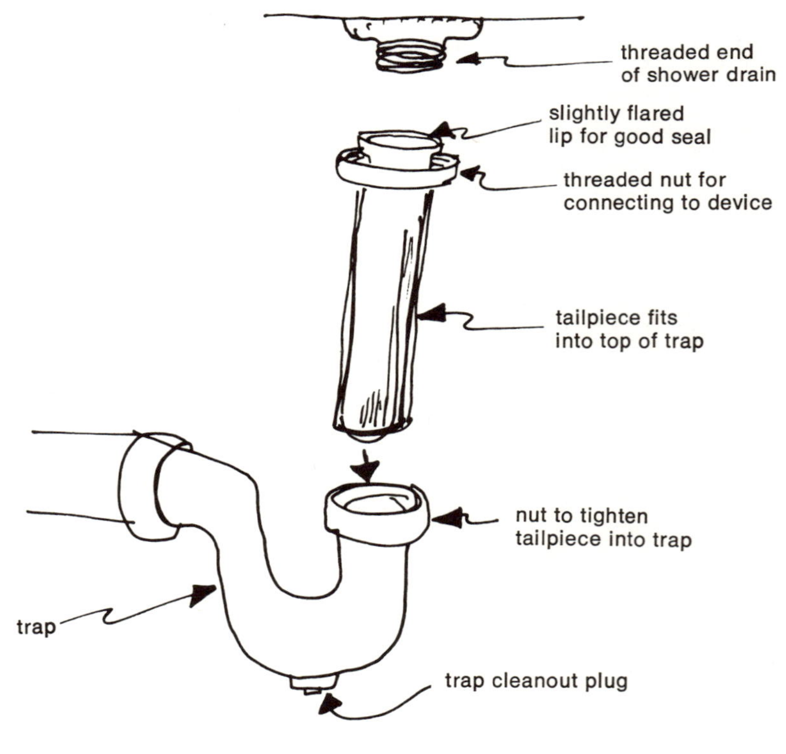

Figure 5.

Hookup of tailpiece from shower base to trap.

To assemble the rest of the shower stall, fit the three sides of the stall together; this must be done in proper sequence, as the metal sides have slots and tabs that must be fitted into one another to give a waterproof seal. There is usually a band of sheet metal around the concrete base over which the metal sides are fitted. Once all the sides are keyed together, you can use the sheet-metal screws provided to attach the whole assembly at the top and at the bottom. Follow the instructions provided with the shower stall, as the exact assembly procedure may vary slightly from manufacturer to manufacturer.

Now you can run your hot- and cold-water supplies to the shower valves and test out the system. Run the shower and check your drainage hookup for any leaks; the tailpiece-to-trap connection must be tightened as strongly as possible to avoid leakage. Once you've made sure there are no leaks, you can enclose the base with some plywood if you like. Attach any covering pieces with screws instead of nails, so you can take the panel out again easily in case problems develop later that require access to the trap and drain fittings.

The above techniques also apply to installing a bathtub on a raised platform, except that you will have to enlarge the dimensions of the base to accommodate the tub. You will also have to construct the base much more solidly to handle the additional weight that a filled bathtub would create. Supports might consist of a network of thick timbers bolted and braced together as strongly as possible. It doesn't hurt to overbuild a little in a situation like this, since underbuilding could be potentially disastrous.

While you are laying in basic drainage, you will probably want to run new water supply lines to any additional devices you have planned—sinks, tubs, darkroom, laundry, and so on. The time is now if you are going to do anything about improving water pressure. If you notice that the flow from the sink is poor in volume and rate of flow, or that the toilet takes several minutes to fill up again after flushing, you may have old, constricted lines that have rusted practically closed. See if you can locate the water riser that supplies your floor. If it is fairly old but of sufficiently large diameter (1″ to 1½″), you can still obtain adequate pressure by installing fresh pipe from the point where the supply reduces down for your floor. You will have to shut off the water supply in the building for a few minutes to expose the reduction point.

Corrosion is usually worst at the fittings—that is, at the Ts and elbows and especially at the reduction point from the riser to your floor. Quite likely, you'll have a ½″ feed for your floor, which with age and corrosion may be reduced to less than ¼″. It's well worth the expense of continuing your plumbing from this point on in noncorrosive copper tubing or threaded brass pipe. Install a length of pipe with a good shutoff valve at a point that will be accessible to you later when you need to shut off the supply to your floor for repair work or modifications. Once you've got the length of pipe and turnoff valve

Enlarged platform sturdily built
and capable of holding the weight of a tub.

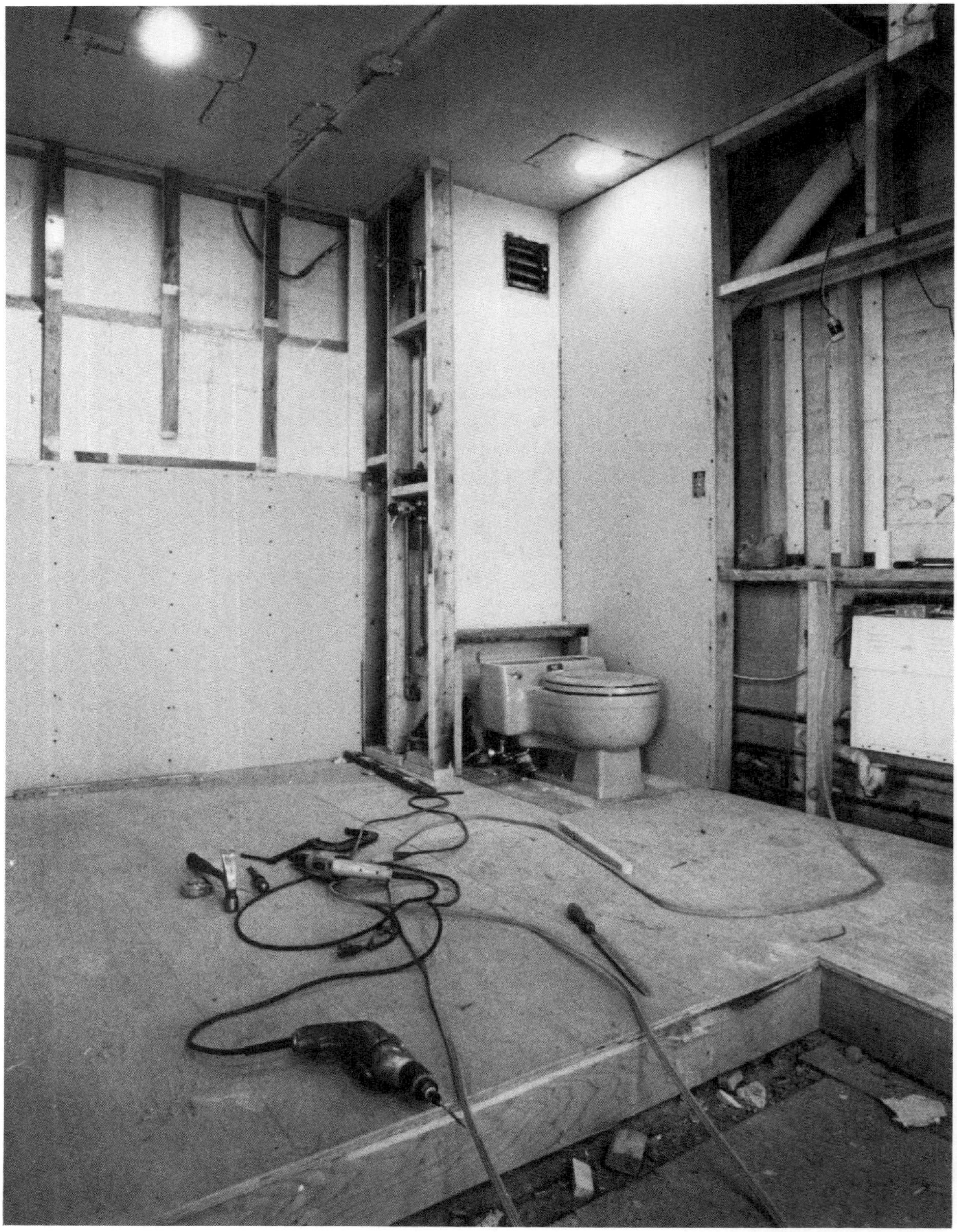

installed, you can turn the water supply for the building back on and continue working at your leisure, without inconveniencing the other tenants. Don't make the mistake of installing small-diameter (½″) galvanized pipe, as it corrodes within five or ten years in most city supply systems and will need to be completely replaced. Large-diameter (¾″ or more) galvanized pipe will last a lot longer, but its greater cost obviates the savings versus using small-diameter copper and brass. If local regulations require the use of threaded pipe, you'll have to get a professional plumber with a threading machine or go out and get a set of taps, a portable stand, some cutting oil, a pipe cutter, and a few wrenches and try to do it yourself. It's not all that difficult, but doing threads by hand can be backbreaking and you might not be up to it.

As mentioned in Chapter 4, you should try to keep all your plumbing runs to a minimum, especially if you are on a limited budget. Metal of any kind is getting increasingly costly, so try to keep your plumbing-related activities reasonably close to your supplies. With careful planning and designing within the space allotted, it is possible to save on materials and still have a good layout.

INSTALLING A HOT-WATER HEATER

While you are doing rough-in plumbing work, you may also want to install a hot-water heater, unless the building has a hot-water supply. Most lofts don't, so you'll have to provide your own. Try to figure your hot-water requirements in advance; if you are going to have only a shower and sink, you can get by with a 30-gallon heater, which is a fairly average size. Putting in a tub, washing machine, and dishwasher will necessitate a 40-to-50-gallon unit, with a high recovery rate if you have a family or large group of people. The bigger the tub, the more water you will need. If you're planning a free-form customized tub or huge group-bathing facility, figure on a 75-to-100-gallon unit with a triple-stage heater—as well as a laundry-sized fuel bill to run it.

Plan on installing your heater in a closet or someplace where it can be easily reached and removed if it breaks down. Gas-fed units will require occasional access in order to light their pilot-thermostat assemblies, so you'll want to allow some elbow room. Remember that gas hot-water heaters must be vented by attaching the correct-size flue pipe to an appropriate opening (through a ceiling and roof, or into an air shaft); otherwise a noxious and potentially harmful vapor will be wafting into your loft. Some old-time lofts were so drafty and leaky that gas hot-water heaters were run unvented for years without death or injury; it's not a safe practice, and venting is well worth it.

If you are in an all-electric area, you'll have to install an electric unit. Electric 220-volt heaters can be installed anywhere, since they require no venting. But don't box them in completely—they can fail and need

This hot water heater is fully accessible through a closet and correctly vented.

replacement. In some cases electric heaters have faster recovery rates than gas-fed units and certainly require less maintenance. There are no pilot lights to worry about, but they do use a lot of electricity.

If you are in a building with steam heat, you may want to make any contemplated modifications to the system at this time, especially if the changes involve running pipe under platforms or elevations. For steam work you may require the services of a professional plumber, as the pipe sizes involved are rather hefty and all the work has to be done in steel pipe with black cast-iron fittings—which is pretty arcane stuff unless you already know about it. Also, if you are working in the winter, you'll have to shut off the building's steam to open the main risers. All this could be dangerous if you don't know what you're doing and open a 2″ steam line in your face by accident. So don't fool around here; unless you feel very competent, get a pro for the work and show him where you want to put the lines, extra radiators, and so forth.

At this time you will want to bring in any supplies for hot and cold water under areas that are going to be raised or platformed over; trying to do this later is just awful work, if not impossible. This also applies to electrical cables, telephone wires, and the like—they should all be installed in the beginning, not later on.

INCREASING WATER PRESSURE

While you are putting in your plumbing, you can correct any deficiencies in your water pressure. Low pressure can be caused by a variety of factors. One of the most common reasons is that the pipe coming off the main riser to your floor has a severely reduced feed capacity because of rust and scale over the passage of time. This can be corrected by going back to the point of origin of your floor supply and installing fresh pipe. In other cases the problem may lie in the main riser itself. Risers can corrode and cut down pressure over the years, especially if they are small in diameter (½″ to ¾″) to begin with. Risers can be replaced, and in many cases the landlord will pay for it. Sometimes even a good-sized riser (1¼″), recently installed, will still have low pressure to your floor. A street-level pressure of 40 pounds is considered to be fairly common; for every 2.3 feet of elevation above the street, the pressure drops a pound. If you are on the fifth floor of a building with 12-foot ceilings, you would have only 20 pounds of pressure to work with. You might find this insufficient for showering or filling a large tub. One solution is to purchase a pump with a pressurized holding tank, very similar to those used in many suburban homes. Basically, the pump pulls the water up the riser, pushing the water into a holding tank, where it builds up pressure; the water is then sent out into the system at say, 50 pounds of pressure for use in a shower. The pump has a cutoff mechanism to keep the pressure between 30 and 50 pounds.

Pumps do a nice job of boosting usable pressure, but there are two drawbacks—cost and noise. A holding tank costs about $185, and a good one-horsepower pump costs about $275 (these are approximate prices). The noise problem is another matter. In suburban or country homes pumps are usually located in a basement or crawl space and are thus almost inaudible to the people in the house. But in a loft a pump is more likely to be placed in a utility closet near the water supply. It can be loud. It is possible to muffle the noise by covering the inside of the closet with sound-absorbent material, such as fiberglass or quilted-foam-rubber insulation sheets. Another solution is to place a switch for the pump in your bathroom and turn it on only when you want increased pressure for a shower or bath. Depending upon your exact situation, you may be able to get away with the existing pressure for your other, nonshower needs. One other problem with a pump setup—when you've got your pump on, your neighbors may have a pressure drop.

CUTTING INTO MAIN DRAINS FOR SHOWERS, BATHS, SINKS, OR TOILETS

Certain circumstances may arise where you may not want to build a platform high off the ground to hook up devices that require drainage. Or you may have a perfectly good drain riser passing through your floor but with no existing fixtures hooked into the drain. If you want to connect a fixture to that drain stack, you will have to cut an opening into the drain. This can, in fact, be done without hiring a plumber to cut out a whole section of the drain and reinstall a section of cast iron with a Y connection. Many loft dwellers don't have the financial resources for such work. The alternative consists of cutting a hole in the cast iron with a hole saw (or a series of small holes, breaking out the pieces) and installing a saddle fitting (see figure 6). The saddle is attached to the main drain with a U-shaped hoop and two nuts that are drawn up tight to the fitting. Saddles come in a variety of configurations—Ts or Ys usually—and are threaded to accept either regular threaded galvanized pipe (in the 1½″ or 2″ sizes for sinks, tubs, and showers) or cast-iron hub adapters for the conventional oakum and hot-lead drainage hookups (used only for a 3″ or 4″ toilet hookup). You can use PVC for a toilet hookup with a PVC hub adapter, as described in Chapter 14 (see page 123).

Before you go ahead with a saddle hookup, make absolutely certain that the drain you are planning to cut into is hooked up in the basement and fully usable. After you cut your opening in the pipe, run a hose into the opening and flush some water down the pipe, making sure the water does in fact drain away. You should make a floor-by-

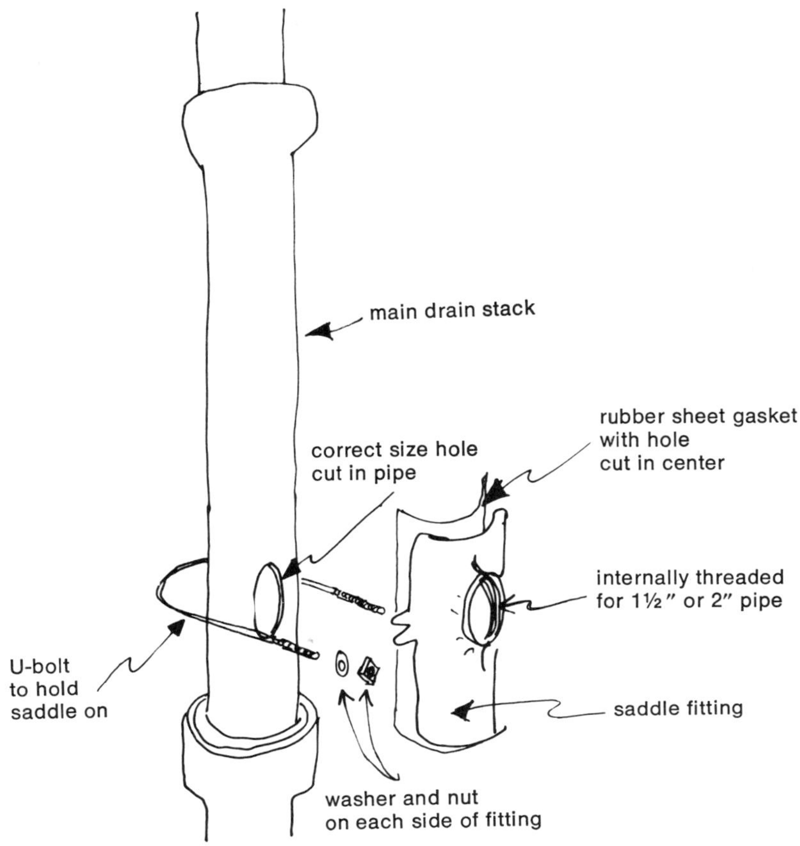

Figure 6.

A saddle fitting can be used to drain into the main stacks without involving costly cast-iron drainage replacement.

floor inspection of the riser to be sure the drain has not rusted through anywhere or been disconnected at some point.

To attach the saddle fitting, lay a sheet of rubber, with the appropriate-sized hole cut out, between the drain riser and the fitting itself. In addition, you should lay in a sausage-shaped roll of A&B plumber's epoxy as well as a bead of epoxy over the edges of the opening to provide as smooth a surface as possible for the flow of waste water down the drain. Allow the entire assembly to dry for at least 24 hours after installation, as the epoxy is water-soluble until it sets properly. If you don't allow it to dry properly, you may wash away your sealant with your first shot of waste water.

With reasonable care in assembly and installation, a saddle fitting can give good service for many years and make possible a whole new layout in your loft space.

.9.
Insulating Techniques and Installing Auxiliary Heating Systems

Very often a loft that has a good steam-heating system during the winter will have one drawback—it goes off at 5:00 P.M. and on weekends. You can get pretty cold after two days without heat, so you may want to install some sort of auxiliary heating system.

Unfortunately, the "energy crunch," or whatever you want to call it, has skyrocketed the cost of electricity to the point where it costs several hundred dollars a month to heat an average surburban home. Many loft spaces are even larger in total square footage than suburban houses; in addition, lofts tend to be a little drafty, thus raising the heat-loss factor even more. All the electrical heating devices available today were invented back in the good old days when electricity was dirt cheap and electric heaters were inexpensive and easy to install.

Alas, the engineering behind heating electrically is poor. In order to heat up a space, you have to use a lot of electricity as compared to the amount of fuel oil that would be required to do the same job if the oil was used directly rather than first being turned into electrical energy and then used to heat up a room.

Even though utility rates vary around the country electric heat costs plenty these days. The highest rates are in the Northeast, with New York City the most expensive at nine cents per kilowatt hour. The cheapest rates are in the Southwest—Dallas is the lowest at three cents per kilowatt hour. If careful insulation is done, along with a sound program of electrical conservation, you can keep bills within reason.

If you have a gas supply, you will probably opt for gas-fed auxiliary space heaters, as they will cost far less to run than electric units. Even though the price of natural gas has shot up tremendously over the past few years, it is still more efficient than electricity. There are a couple of drawbacks to gas heaters: they need venting of the combusted by-products, in the same manner as gas-fed hot-water heaters, and they need pilot lights and the attendant maintenance necessary for such devices. Electric heaters usually require little maintenance and from a strictly practical point of view are safer to operate. All gas heaters sold today have virtually foolproof pilot-safety mechanisms that eliminate the possibility of an unlit jet filling a room with gas. If the pilot light is not on or not working, no gas will flow through the safety valve into the main burners, where it would escape into the air.

All the following insulating techniques apply equally well to gas and electric heating.

First, try to seal up every possible entrance point for cold air. Check old window frames for cracks and open spaces to the outside. Narrow cracks can be filled with caulk, applied with a caulking gun. Use a good exterior-grade caulk and remove any loose dirt or peeled paint before caulking. Larger gaps can be stuffed with oakum (available from plumbing supply houses); extreme cases might call for a filler of cement, which can be a ready-mix material such as Sacrete. Mix it to a stiff consistency and trowel it into the opening, making sure to wet down the brick with some water first for a good bind.

You may even wish to install storm windows fitted with double-thickness glass (thin glass is a terrible insulator) or, even better, dual-thickness glass with an air pocket between the sheets; this is often called Thermopane. Sometimes it is cheaper and easier to install an additional set of windows than to replace all your windows and frames. Check prices from several different outfits first, and ask to see samples of all the materials used. Be aware that there are several different thicknesses of aluminum used in storm windows; obviously, the thicker metal is stronger, lasts longer, and is more resistant to corrosion. Anodized finishes further protect the raw metal. In general, use common sense—a very low bid can mean trouble in the form of substandard materials, poor or sloppy installation, and other problems. Avoid paying in full until a job is completely finished to your satisfaction—don't let them say "I'll have my man stop by next week and replace those broken pieces" and then take your money with them. Hold out until they finish; you'll be surprised how fast they correct or "find" whatever piece is missing or defective.

If you can't afford any of the above, you can still save a lot of heat by covering your windows with large sheets of plastic. To seal up the space between the plastic and the wall, staple the sheet down and cover the seam with gaffer's tape (also called plumber's tape or duct tape). This will keep out much of the draft.

Next, check your entrance doors for any gaps or spaces; try to avoid opening hallway doors unnecessarily, as every time you do you lose heat. If you have a truly immense space, you may simply want to partition off a small living space and insulate and heat only that which you need for comfortable living. In addition to putting up walls and laying up thick glass wool insulation, you should drop your ceiling if it's over 9 feet high. You'd be amazed at the amount of heat that sits up on a ceiling.

If you are going to be installing gas heaters for your main or auxiliary source of heat, you should also run the necessary gas lines at this time. Depending on the size of your space, you'll probably need at least a ¾" supply pipe. Remember that all gas work should be done in "black" pipe and fittings, for proper compliance with code. Floor-mounted heaters are relatively easy to hook up; run the pipe over from the gas supply pipe, using a shutoff valve and union fitting for the final connection to your heater. Connect the gas line by means of an elbow or flexible gas connector to the piece of threaded pipe at the back of the heater. Each piece is put on one after the other, as in conventional threaded-steel water pipe.

To vent the unit, attach a vent elbow (3" or 4" in diameter) to the short vent nipple protruding from the back of the heater (see figure 7). Venting material is available in either 2-foot or 10-foot lengths. Run venting to an appropriate spot, usually a chimney opening or the top of a window. If you have a long run, use the 10-foot lengths so that you don't have to screw or rivet together several short pieces and you prevent any possible sagging. Remember to pitch the run upward from the heater, using as few bends and elbows as possible.

Ceiling-hung units are a little more trouble, but if you are trying to heat a large space they are the only practical things to use. You will have to locate some of the ceiling beams directly above the unit and sink some heavyweight eye hooks directly into the beams. Then, using steel-threaded rod, some pipe Ts, and other fittings, make up the necessary support pieces for the heater. The usual practice is to utilize two unions for the final hookup in midair. You'll need some help getting the unit up to the right height. Place the heater on a stepladder and have a few people hold it up while you screw the two unions together. Once

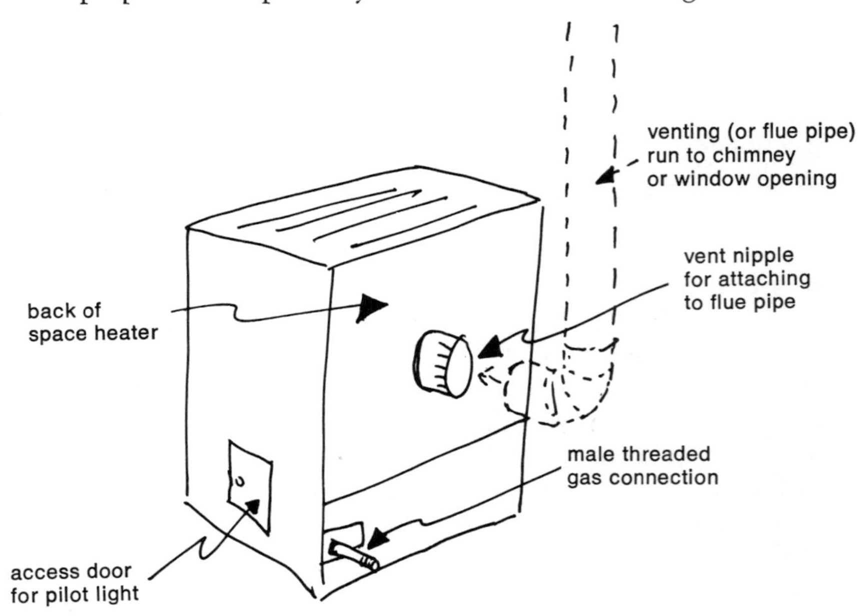

Figure 7.

Venting (or flue pipe) is attached to the nipple at the back of the heater by means of a vent elbow.

Ceiling-hung gas heaters
are the best way to heat large areas.

the heater is in midair, you can make the gas hookup, as well as installing the 110-volt electrical line for the fan and an additional pair of wires to a wall-type thermostat. Some gas heaters can have their thermostats hooked up with bell wire, as they are on a low-voltage system; others utilize a line-voltage thermostat setup, which will require a pair of No. 12 or No. 14 wires.

If gas is not available in your building and you want a standby auxiliary heating system, you'll have to go for electric heat. During the partitioning and wiring phase (see Chapter 11) you must run the appropriate 220-volt supply line to each place where you want a heater. As to the number and size of units to use, only you can work that out, with the help perhaps of a friendly electrical supply house salesman—or a heating engineer, if you know one. Suffice it to say, your requirements will vary according to the part of the country you live in, the construction of the building, the type and size windows you have, whether or not any insulation was ever used in the building, and the exposures of your walls and windows. Baseboard heaters are available in lengths up to 10 or 15 feet. The longer ones require more juice. Be careful not to obstruct the units with furniture or draperies; these objects might start to smolder and eventually burn if left unattended too long. The units are usually screwed to wall baseboards a couple of inches off the floor. The wiring is then brought through a knockout in the back (for use on sheetrocked walls) or through the sides (for surface-mounted electricity, such as EMT tubing). You will save money in the future if you install a separate thermostat for each room, so you can turn off unused areas when not occupied. Since electric heat comes on immediately, if you have a reasonably well-insulated space you can turn off the heat when the room is not in use; the space will probably not drop more than 20 degrees in temperature.

.10. Floors: Sanding, Refinishing, and Installation

Along with the manifest benefits of loft living, one of the most visible and appreciated features has been beautiful old wood flooring. Wood floors, rated for industrial loads, are very often found in loft spaces. Although a floor may seem in poor condition, the original glowing colors of the raw wood can often be restored. Some floors are real gems, made from old and rare woods that are no longer available in the necessary quantities for flooring. Others are made from more common woods, such as pine, oak, and cherry, and have worn down with age and use—and abuse too, due to years of leaky machinery from which oil stains built up. The oil stains on these marginal floors cannot be sanded out completely, and it is sometimes better to paint the floors with deck enamel rather than attempt refinishing.

Many loft floors can be brought to some reasonable semblance of finish and luster by means of sanding and refinishing. The best way to get a decent finish is by renting a professional floor-sanding machine. Trying to sand down a large floor with a small, hand-held orbital or belt sander, which many do-it-yourselfers have, is madness. Not only will you burn out the tool motor very quickly, you may also break your back! The right tool for floors is a professional machine that weighs well over 200 pounds and costs about $800. It's not the sort of machine you'll ever use again, so unless you are going to go into the floor-sanding business, get your Yellow Pages (a fabulous guide for tracking down all sorts of things) and look under "Floor Machines–Renting," "Floor Refinishing and Resurfacing," and the like. Check out some of the rental places for the best prices and weekend specials. Very often if you pick up your machine on Friday night and return it on Monday, you'll be charged only for one day's rental; and if you're strong enough, you can use the machine Friday evening, all day Saturday and Sunday, and do your whole space in that time. Since the machines are very heavy, you'll need a few strong friends to help you get the equipment up the stairs—if you're not lucky enough to live in a building with an elevator.

Most of the places that rent floor-sanding equipment will show you how to install the paper on the drums as well as tell you how much paper you will need for the job. Paper is sold by the sheet, and most rental places will take back unused sheets for credit or refund. They also sell the various finishes—varnishes, shellacs, polyurethanes, and the like—and can advise you as to durability and application methods. Despite its high cost, polyurethane is a big favorite because of its durability and luster. If you want, you can add wax on top of the polyurethane and buff up the floor every once in a while.

In addition to the sanding machine, you'll need an edging sander to get into corners and edges; it's hand-held, but in name only—in using it, you work on your knees. The edger rolls on two little wheels; you tilt it down to make contact with the floor and tilt it up to stop. You'll need some sandpaper rolls for the big machine and sandpaper discs for the edger; the grade of paper you use will depend on what condition the floors are in. For taking off paint you'll need a lot of coarse-grit paper to get down to raw wood; then some medium paper to get it all a bit smoother; and finally some fine paper to give the floor a nice finish. If you omit the fine paper because you are planning to polyurethane the floors, keep in mind that you will need a lot more polyurethane to do the job. Rough-textured wood is much more absorbent, and additional coats will be required to obtain a high degree of finish.

Sanding floors is backbreaking and time-consuming work; the air is hard to breathe—even with a good dust mask you'll wind up with dust in your eyes, ears, and mouth (although you can minimize this by taking frequent breaks and working with good ventilation). Therefore you may prefer to call in professional floor sanders to do the entire job for you—sanding as well as refinishing. It will certainly cost you more than doing it yourself, but not always that much more, considering the cost of equipment rental and transportation, paper, and finishing materials.

If you don't have a wood floor in your loft, it is possible to install one yourself or, even better, to have a flooring outfit come in and lay one down. In fact, in some cases you may wind up having to install new flooring over an existing wood floor. For example, if there is a tilt or list to the floor, you will have to correct it by shimming up the depressed area and laying new flooring over it. Or an old floor may simply be too far gone for your taste—the planks may be rotted or split severely or worn down to the point where the knots are raised. A really severe slope or tilt may indicate a structural decay—cracked or rotten beams—that should be investigated. If in doubt, try to get a structural engineer's opinion as to the soundness of the floor. If your landlord has any interest in keeping his building intact, he may effect the necessary repairs himself in order to keep his building from falling down. As mentioned above, a lot can be done by simply laying down a shimmed-up underfloor (of, say, ¾″ plywood) and then installing a hardwood floor (maple or oak) over it. The new floor should last for another 100 years!

It will cost you quite a bit of money to lay down a new wood floor—ranging from $1 to $2 a square foot even if you do it yourself. But there are some advantages to letting pros do the job. They'll do it much faster and better than most people can do themselves. When nailing a floor down, professional floor layers use a special jig that drives each nail in at a precise angle without splitting the wood; a good flooring man can drive nails in at a pretty fast pace. Another reason for using professional flooring companies is that they have large inventories of woods, and since they buy in bulk they usually get a good price, which is in some respects passed on to you. If you go out and try to buy flooring, you'll pay a fairly high price for the materials to begin with, in addition to having to lay it down yourself.

What do you do if you have concrete floors and desire to have wood floors? There are at least two choices: you can literally glue wood flooring to the concrete, using a tarlike mastic, or you can lay down wooden sleepers to duplicate the effect of floor joists and nail the strips of flooring to them (see figure 8). This latter approach, although costly, is far superior to gluing the wood down with mastic, especially if you are in a damp location or have leaky roofs or windows that will let in moisture when it rains. If moisture gets in under wood floors and has no place to go (concrete being waterproof), it will be absorbed into the wood above, swelling the wood strips by as much as ⅛″ laterally for every 3″-wide strip. Clearly, over the width of a loft the floor could increase in size laterally to a rather large degree! Unless the swollen flooring pushes down the two walls it is expanding against, it will inevitably rise in the middle, creating a bulge or camel's hump, an upside-down boat pushing up through your floor, or some other such ghastly effect. This can be a nightmare in a very real sense. Recently one loft dweller we know heard creaking and rending of wood during the night—something reminiscent of sailing in an old wooden boat. Upon waking the next day she discovered a large hump in the middle

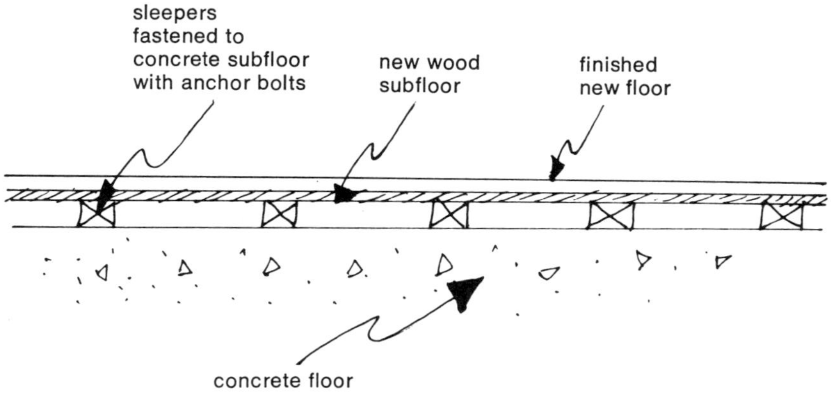

Figure 8.

The best method for installing a wood floor over concrete is to lay down sleepers, which have been secured to the concrete subfloor with a stud driver or lead anchor bolts.

New wood floor in the process of being laid out.

Nailing floor down using nailing jig.

of the floor. A rainstorm on an extremely humid day had caused the problem. The rain had soaked through the brick exterior walls and cracks in the window frames and down onto the concrete underfloor. The trapped moisture did the rest. The situation could have been prevented by waterproofing the walls and sealing the window frames tightly to prevent moisture from coming in. Also, had the wood floor been placed over sleepers, any trapped moisture would have had some air space in which to evaporate—between the wood and the concrete—and the wood, securely fastened to the sleepers, would not have been able to bubble up.

If you are planning on laying new flooring, you should do it before any partitioning work, to give you something to lay your stud plates down on and also to make it easier to install the flooring in one large area, rather than having to work inside already existing structures later on. You should hold off on the sanding and finishing until you have at least painted your ceiling, if not the walls. Then any paint that may have been splattered from the ceiling can be sanded away before refinishing the floor. If you want to be rid of the dust that sanding creates, you can sand and finish your floors immediately after installation and then protect them with heavy special-purpose paper, taping it down to the floor and along the edges of the walls. If you're lucky and don't tear any holes in the paper, you can pull up the paper once all your construction is finished, buff up the floors, and all should be well. If you're painting your floors, you should wait until you've finished painting the ceilings and walls.

.11. Partitioning and Wiring

PARTITIONING MATERIALS

Once you have your basic plumbing and new flooring in, you are ready to put up essential partitions. Although it's nice to keep partitioning to a minimum in lofts, some may be necessary, especially for bedrooms, bathrooms, storage, or any other areas that require physical isolation from surrounding open space. There are two different ways of approaching the division of space: you can completely enclose a room, or you can put up a partition that does not extend to the ceiling. There's no rule that says you have to install partitions all the way to the ceiling; in fact, for some purposes it may be superior both from an esthetic and practical point of view not to.

There are certain basic tenets of construction that apply to all types of partitions. First a discussion of the various materials involved.

Wallboard or Sheetrock

With the post-World War II advent of wallboard (or sheetrock)—a synthetic panel of hardboard, gypsum, asbestos, plywood, or other material—the time-consuming task of wet-plastering walls has been tremendously simplified. Before the invention and widespread distribution of this revolutionary product, walls were traditionally covered with lath and plaster, requiring a skilled craftsman literally to nail up sheets of metal lath to the studs and then completely cover them with wet-mix plaster. With wallboard, all you have to do is stud out the area to be divided and nail up the wallboard. You may still have to call in a plasterer to seal any joints between the sections of the board with joint compound (unless you are proficient at this), but the total amount of labor needed is much less than with the old method.

In terms of absolute strength and soundproofing, the lath-and-plaster method is somewhat superior and is still used in institutional or luxury high-rise construction. However, there is a tremendous amount of skill involved in plastering smooth walls, and very few people are trained to do this work today. From the point of view of cost, wood studding and sheetrock is the cheapest way to go. You can save further by not taping joints at all or just using gummed tape or masking tape. But the taping, if properly done, can make your walls look as if they were perfectly plastered. What you do is a matter of taste, ability, and money. Remember too that you're not restricted to sheetrock for covering walls or partitions. Plywood can be nailed to conventional studding; or you can staple wide rolls of no-seam paper over a simple wood frame for a quick or temporary division of space.

Most people tend to use the conventional wood studding and sheetrock. Sheetrock comes in several different thicknesses. The easiest to score, break, and install is ⅜″ thick. If ease of installation and cost are primary, this will be the best kind to use. Certain fire and zoning codes prohibit the use of ⅜″ sheetrock, so make sure to check before you buy and install it. The ½″ size is considerably heavier and harder to carry and work with, but it usually meets code requirements. The thickest sheetrock is ⅝″ and is often code-required on staircases, as well as interiors in some localities. Most thicknesses of sheetrock come in a variety of sizes, the most common being 4′ × 8′ and 4′ × 6′. If you are very sure of yourself and have some strong and able co-workers, you may want to buy 4′ × 10′ or even 4′ × 12′ pieces. These larger slabs weigh a lot and are more easily broken; only those with professional experience will want to play around with them. The 4′ × 8′s and even 4′ × 6′s are handful enough for most do-it-yourselfers.

There is also a special kind of water-resistant sheetrock for use in bathrooms or any damp locations. If you are planning to do any tiling, it is essential to use this type to make sure your tiles adhere properly and do not fall off within a short period of time. Water-resistant sheetrock can be distinguished from the regular dry-wall type by its light green color and the letters M/R (moisture resistant) printed on the edges.

Studs

In addition to a choice of sheetrock thicknesses and sizes, you have two options on the type of material used for studding—wood or metal.

For most applications, wood studding is the best and easiest to use. The most common size is the 2″ × 4″ stud, available in lengths ranging from 8 to 20 feet. There are several different grades of this size wood. The grading has to do with the amount the wood has dried out and the straightness of the piece. Kiln-dried wood is usually the best, straightest, and most expensive. If you're on a low budget and just need some quick, rough work done, you can use a cheaper grade of construction lumber that may be a little green and a bit twisted or warped but adequate for your requirements. If you want to make sure of not getting crooked wood, go down to the lumberyard and pick the pieces out yourself.

In some cases you may not even need 2″ × 4″s. If your wall does not have to be immensely strong or you are dividing a space with a low ceiling, you may be able to use 2″ × 3″s, which are slightly smaller and therefore a little cheaper.

In certain areas fire codes may specify the use of metal studs for any partitioning within a certain type of building. Metal studding is used in a lot of office construction today, and it does have superior fire-retardant qualities, as well as being a little quicker to erect and install. By its very nature metal studding is not as solid as wood and feels a little flimsier, particularly if used for free-standing walls. The ease of assembly (with a crimping tool instead of hammer and nails) is offset a little by the need to invest in a screwgun (about $70 to $100), which drills a special Phillips-head screw through the sheetrock and the metal, fastening the two together. With the screwgun, the work goes quite quickly; the tip of the tool is magnetized so that the screw stays in place until it is driven in. In addition, the tool is adjustable for setting the depth of the screw exactly as you need it. The taping and plastering of the wall will go much easier later on if nails or screws are set in properly. The fastener should be recessed, but not so deeply as to tear through the paper covering of the board. If it is torn through, the holding strength of the nail or screw is diminished considerably. Once you've got the correct setting on the screwgun, it'll drive screws to the same depth all day long. The adjustment will have to be varied depending on the type and thickness of sheetrock as well as the material used for studs. If you have to use metal studding and don't want to invest in a screwgun, you can rent one for a few dollars a day. If you are planning to do a huge amount of sheetrock and metal-stud work, it will pay to buy the tool. Good-quality screwguns can also be used for wood screws as well as various types of metal-fastening screws, such as Tek-drives. So it's not just a one-time-use tool, especially if you are planning to do a lot of work yourself.

ERECTING FRAMES FOR PARTITIONS

Erecting a wall or partition is accomplished in one of two ways, depending on whether you use wood or metal studding. For wood studding, begin your framework by laying a chalk line on the floor where you want the wall to be. To "snap" a chalk line, you will need a helper, or a nail driven into the floor, to hold one end of the line (usually impregnated with either blue or red chalk); stretch the line tight at your end and at the same time draw the line up straight a few inches, letting go quickly. A straight line of chalk should remain on the floor after the string is snapped. Using this line as one edge of the wall, lay down a sole plate of the same size wood you will be using for the studs—probably a 2″ × 4″ or a 2″ × 3″ if you are doing a small area only. If you have a wood floor underneath, nail the plate down with some tenpenny nails; if you are going into a concrete floor, use spiral masonry nails or cut nails (shorter nails are best for hard concrete). These nails can shatter with impact, so it's a good idea to have eye protection when using them, as well as a heavy hammer. Some concrete floors cannot be penetrated with nails at all. In this case you'll have to abandon nailing and go for lead anchors and lag bolts or machine screws. You'll need to rent an impact hammer drill for making the holes in the concrete for the lead anchors. Drill holes in the wood sole plate for the bolts and lay the plate down on the line; mark the holes on the floor for your anchors. Then drill the holes for the anchors and place them in the holes. Finally, lay the plate down, line up the holes, and insert all the bolts. If you are using lag bolts, try to get them with hex heads so you can use a ratchet wrench—without a ratchet, you'll have to use an open-end wrench or pliers, which may slip as you start to round off the edges of the head of the bolt. A useful tip: Place a big, flat washer over the wood; otherwise the head will crush into the wood and disappear, leaving you without anything to grip onto with a wrench.

Once you have the sole plate down, begin to mark off some 16″ center lines for the placement of vertical studs. Exact center lines are really critical only at the 48″ point, and multiples thereof, because this is where you will have to attach two pieces of sheetrock; you'll have only the thickness of the stud to do it in, which in most cases will be 1½″ (the new 2″ × 4″s are now only 1½″ × 3½″). If you are trying to save money, you can economize by studding every 24 inches instead of every 16. If you are thinking of installing any insulation between the studs, bear in mind that it comes only in 16″-wide rolls. If you have 24″ spacing, you'll have a gap in each row of insulation, and cutting lengthwise down a roll of fiberglass insulation is not much fun.

At this point you should cut a piece of wood the same length as the floor plate to be used for the ceiling plate. Check the ceiling above the site of the intended wall to make sure that it is clear of pipes or electrical cables. If there is an obstruction, notch out the back of the ceiling plate so it will lie flat against the ceiling. If the obstruction is greater than half the thickness of the plate, you'll have to have a break at that point and continue on with a separate plate. Next, cut two studs, one for each end of the wall, and with a helper raise the ceiling plate into place, using the two studs as levers (see figure 9). Stand on a ladder at one end to guide the plate and stud into its proper position; have your helper stand on a ladder at the other end so that the plate can go in evenly at both ends simultaneously. Be sure to measure the two studs carefully, allowing for the thickness of the sole and ceiling plates (3″ total) and a little bit more for a good wedge fit. With a good fit you can make your adjustments for plumb and trueness without the whole mess falling down on your head as you hit the ceiling plate lightly with a hammer. Once you've checked for plumb, nail the plate into the ceiling

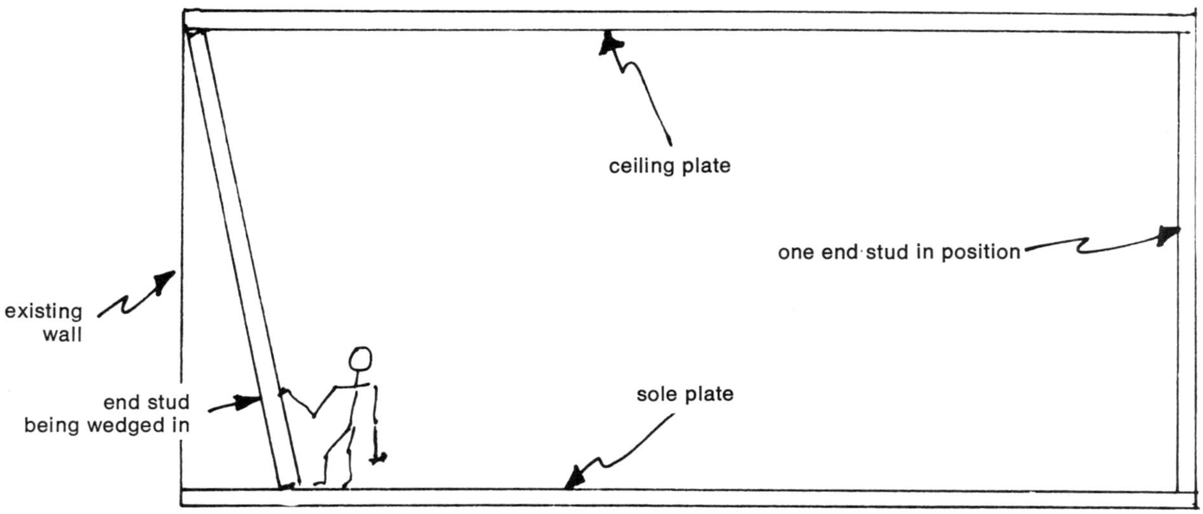

Figure 9.

To raise the ceiling plate into place, use the two end studs as levers.

and fasten the studs at each end of the existing walls. If your ceiling is hollow, you may want to use toggle bolts. For nailing into most brick walls, cut nails or concrete nails are OK. The length of the cut nail you use will depend on the soundness of the material you are going into. Old bricks are soft and require longer nails to ensure firm hold; new bricks or masonry are harder, and shorter nails can be used.

Once you've erected all your framing, short horizontal pieces of wood, called fire stops, should be nailed between the studs in a staggered pattern as shown in the photograph on page 96.

Once you've got the two end studs in with the ceiling plate, you can fill in the studs in between. Unless you have a very flat, even ceiling, you will not be able to cut all the studs to the same length. For a good, solid wall, you should cut each stud as you go along. Toenail the stud at the bottom (that is, drive a nail in diagonally so it goes through the

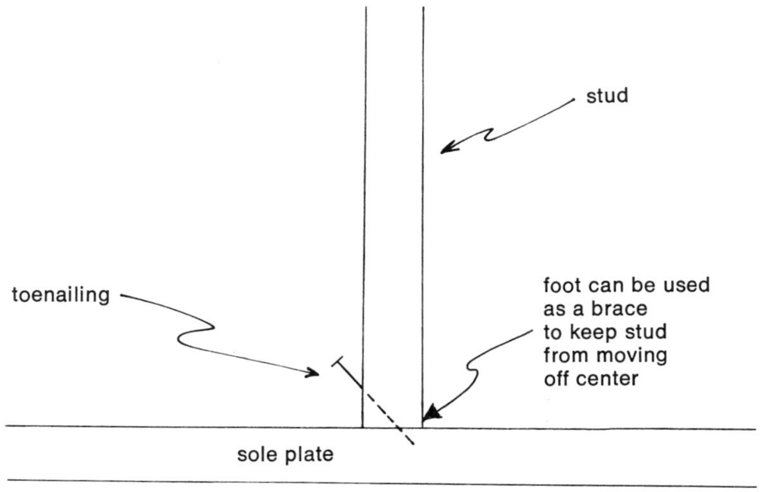

Figure 10.

To toenail the stud into the sole plate, drive the nail diagonally through the base of the stud into the sole plate. To prevent the stud from shifting off the center line, place your foot next to the stud and brace against it while you hammer.

Bathroom framed out and ready for sheetrocking.
Note fire stops and wiring in place.

surface of the stud into the sole plate) and center the piece over the line you have marked. If you have a tight fit to begin with, you should be able to drive a nail in without moving the stud from position. If you don't have a good fit, use shingles for shims. Just slide each shingle in until it gets nice and tight; then nail down the stud and break off the excess shingle with your hammer. Once you've got the bottom located, you'll have to go up on a ladder, level up the stud, and toe-nail it into the ceiling plate.

The same techniques described above apply to metal studding also except that with the metal-stud system the top and bottom plates are metal U-shaped channel tracks. Metal studs have to be squeezed in at an angle and are held in by friction and either crimped or screwed together more permanently later on. You need not attach the stud to the metal track at all on straight spans; rather, after each piece of sheetrock is laid up, align the top and bottom of the floating stud with the edges of the sheetrock, thus bisecting the thickness of the stud and leaving yourself enough room to screw on the next piece of sheetrock.

Doors

If you are planning any door openings in your wall, remember to leave a space on the floor to allow for the door frame. By far the easiest approach to doors is to buy prehung units, which consist of a door hinged to a frame that is ready to be nailed into the prepared opening. This eliminates a lot of detail work for the do-it-yourselfer who may not be terrifically versed in the mysteries of hanging doors. Since the door is already equipped with factory-mortised hinges, it is almost guaranteed to align and close properly, as long as it is correctly installed in the opening. In addition, the trim molding is already installed and the holes for a lockset have been predrilled in perfect alignment at the factory. Installing locksets from scratch calls for a fair degree of skill and care, as well as plenty of time. Although prehung doors cost more money than plain door hinges and moldings, the time and energy saved can be worth it, especially when the final product looks as though it was done by a cabinet maker and not just a weekend loft renovator. If you are a cabinet maker, then by all means build your own doors and door frames.

If you do go for prehung door units, install them as soon as the studding has gone up and before sheetrocking is started, in order to be able to bring the sheetrock right up to the finished edge of the door frame. When you order prehung doors, make sure you get the correct jamb thickness for your wall; allow for the thicknesses of the stud and the two pieces of covering material. For example, if you are using wood 2″ × 4″s and sheetrock, figure 3½″ for the sole plate and ½″ for each side of sheetrock, for a total of 4½″ wall thickness, and thus a 4½″ door-frame jamb.

DROPPING CEILINGS FOR EXHAUST VENTING AND RECESSED LIGHTING

One of the reasons architects and builders put kitchens and bathrooms back to back is that they share a common set of needs in terms of plumbing, water supply, and drainage. Another common need is venting, or the exhausting of stale or repugnant odors from each of the respective chambers. If you have an openable window in each of these rooms, matters can be greatly helped. If you don't, you can install an exhaust fan through the wall, the roof, or a common shaftway that runs through the building. Some municipal codes specify a separate shaft for bathrooms and kitchens; others allow dual-purpose usage of a single shaftway.

In order to get to an existing shaftway, you may have to run some square or rectangular sheet-metal ductwork across a kitchen or bathroom ceiling. If you wish, you can readily conceal the duct, as well as any pipes and wiring that might be up there, by dropping the ceiling a foot or two. A lot of loft spaces have tall ceilings, which may not make for the cosiest of bathrooms. Dropping the ceiling will be esthetically pleasing, even if there is nothing you want to conceal. Also, you'll appreciate the smaller room when trying to take a hot both on a cold winter day.

One way to finish off a dropped ceiling is to install a gridwork of recessed lights, known in the trade as "high hats." When finally installed with their trims, these lights have a neat architectural look that adds a built-in touch to a loft. They are best installed in new construction when you have full access to the ceiling beams and you can run all the necessary electrical cable from one light to another.

The installation of framework for a dropped ceiling is not that hard. Assuming you are spanning a space between two walls, you lay up two horizontal strips of wood on each wall with either long cut nails or, even better, lead anchors in the brick and lag bolts to hold up the retaining strips. Once the two support pieces are up, you can nail in the cross pieces. The cross pieces need not be supported in the center if the span is less than 8 or 9 feet across. For a wider span, you'll need a few pieces of support wire fastened to some sort of supports in the old ceiling, especially if you're going to add lights. Once the frames and housings for the lights are in, you are ready to cover your new ceiling with sheetrock or whatever you wish. Note: Before the ceiling covering goes up, check all lighting or electrical connections you have made in the false ceiling area.

Sheetrock is by no means the only good covering for false ceilings. Tongue-and-groove pine boards can lend a very nice feeling to a bathroom or kitchen. If you use pine boards in a bathroom, you should

A dropped ceiling is a good place to conceal plumbing, wiring, and lights. On the left is a closeup of a wired, installed high hat. On the right, the last of a group of high hats is about to be put into place.

Tongue-and-groove pine boards give the ceiling and walls of this small bathroom a country feeling.

apply several thick coats of polyurethane as a waterproofing agent; you may, in fact, want to do the same in a kitchen to make it easy to clean the surfaces of grease and dirt buildups. If you have access to it, old barn siding is a traditional favorite for many applications in renovation work.

RUNNING ELECTRICAL WIRING

At this point you should run any electrical wiring within the wall (or dropped ceiling) for new outlets. Depending on local codes, you may use either BX or Romex cable and metal utility boxes (with straps for attachment to the studs) for switches and outlets. Now is the time to do any electrical work of this nature, as it is hard to conceal work later on without a great deal of wasted time and labor.

When figuring out your circuitry, bear a few things in mind. Auxiliary baseboard electric heaters will need special circuits, the number of which will depend on your requirements and physical situation. Also, if you are planning on air conditioners for a loft, they will probably require separate 220-volt circuits consisting of three pieces of No. 10 or No. 12 wire. The wires should not be crammed into too small a piece of tubing—if you use the next-size tubing, you'll have an easier time pulling the wires through. You'll also need a separate circuit for your refrigerator, as well as for a washer/dryer, dishwasher, or freezer unit. If you have lighting circuits doubled up with appliances, you'll have dimming lights each time the compressors cut in.

An important note: *Always* turn off the power to the section of your floor where you are doing electrical work.

Running electrical cables within new walls is relatively simple if you have some dexterity as well as an elementary understanding of the principles of electricity. First, install your outlet and switch boxes (sometimes called gem boxes) with the brackets provided for nailing to your studs. If you are working with metal studs, you'll need to use special steel clips to attach the wiring boxes; or you can use a drill and sheet-metal screws to do the job. Once you've got your boxes in, run the cable from one box to another. To do this inside wood-stud walls, you'll need to drill a hole about an inch in diameter through each stud. An electric drill with a wood-boring bit will make this work go fairly quickly, although a brace and bit can be used too. To cut the metal sheathing on BX cable without cutting the wires inside, make a diagonal cut with a good, sharp hacksaw blade, and take care in the last few strokes—cut just deep enough to be able to twist the excess metal sheathing off, thus revealing the wires so you can make the connections. Be sure to use the fiber or plastic insulating sleeves in the end of the BX to prevent the wires from chafing against the sharp edge of the BX sheathing and shorting out. BX cable is easily installed in

Gem box nailed to stud and BX wire connected.

metal-stud walls because the studs are perforated every few inches. Run the wires into the boxes, leaving enough extra wire (6″ to 10″ at least) to enable you to wire up the switches and receptacles. Do not actually connect your devices at this point, as they will interfere with the proper installation of the sheetrock. By not putting them in now, you will be able to get a neater fit on the holes you must cut in the sheetrock for the boxes; and when you install the switches and outlets later on, you'll be able to rest the plaster ears of the boxes against the edge of the sheetrock. Otherwise you'll have to cut a hole in the sheetrock large enough to clear the box *and* the fittings, and you may wind up with a hole larger than the size of the standard cover plate—which will necessitate additional tedious plastering to conceal the gap between the plate and the opening.

If you don't feel knowledgeable or competent enough to do the job, you should hire someone to do the wiring in the walls for you. If you decide to call in a pro at this time, you should also give some thought to the rest of the wiring for the loft—new outlets on existing walls and wiring on the ceiling for lighting tracks or supplies to other parts of the loft. Professional electricians figure time getting to and from a job in their price quotes, so it will pay you to give them as much as possible to do at one time—consistent with your construction progress, of course. If you are doing the work yourself, you can complete the remaining wiring later on.

LAYING UP SHEETROCK

With electrical work within your raw walls completed, you can begin to sheetrock over these areas. Do one side at a time, starting from the bottom and working up. Try to keep the sheets square to your studding for proper alignment. If the floor is on a bit of a slope and you lay a sheet on edge, flush with the door, your sides will not be vertical, or aligned with the studs. It is best to pick up one end of the sheet, align it properly with the studs, and then nail or screw it into place. Incidentally, a screwgun can be used with wood studs and wallboard screws—it does a nice job too! (See photographs on pages 104–105.)

As much as you try to set up the construction of walls with perfect 48″ centers, there comes a point when you have to cut a sheet or even a multitude of small pieces for certain jobs. To cut sheetrock, mark out your lines with a pencil (or chalk line for full-length cuts) and then, using a new blade in a utility knife or matte knife, cut along your lines. Don't bother to cut deeply—all you need to do is score the lines. To separate the two pieces, simply exert some pressure on the edge of the piece to be cut away; the sheet should break cleanly along the scored line. Since the thicker boards take more effort to break, you may want to make a slightly deeper cut with the knife to make it easier. Change blades often to make your cutting easier; plaster is very bad on steel blades and dulls them quite quickly. Once the break is made, bend the piece back a little and use the knife to cut through the cardboard backing of the sheetrock. Some care must be taken when cracking off long, skinny pieces from a sheet—if you try to exert pressure from only one end, the piece you need will crack in the middle. Exert your pressure from the middle and, if necessary, work toward each end separately, a little at a time. After a while you will find the scoring and cracking of sheetrock to be quite simple, and a little bit of patience and care will pay off in the long run.

Cutting out holes within a piece of sheetrock—say, for an outlet box in the wall—can be done carefully with a knife blade. A far easier method for the average loft builder is to use an ordinary saber saw. Special wallboard blades are available for this, but any old coarse-toothed blade will do the job fine. Make a plunge cut with the saber saw and then follow the lines for the box opening you have drawn on the piece of sheetrock. You can use a knife if you don't have a saber saw, but you'll have to cut all the way through the board, as there is no way to score and break it off. There are special gadgets on the market for cutting outlet holes in sheetrock, but they cost money and fulfill no other use. Most people working on lofts already have at least a saber saw, so there's no need to buy such a highly specialized tool.

To cut out a special shape (see figure 11) from a piece of sheetrock,

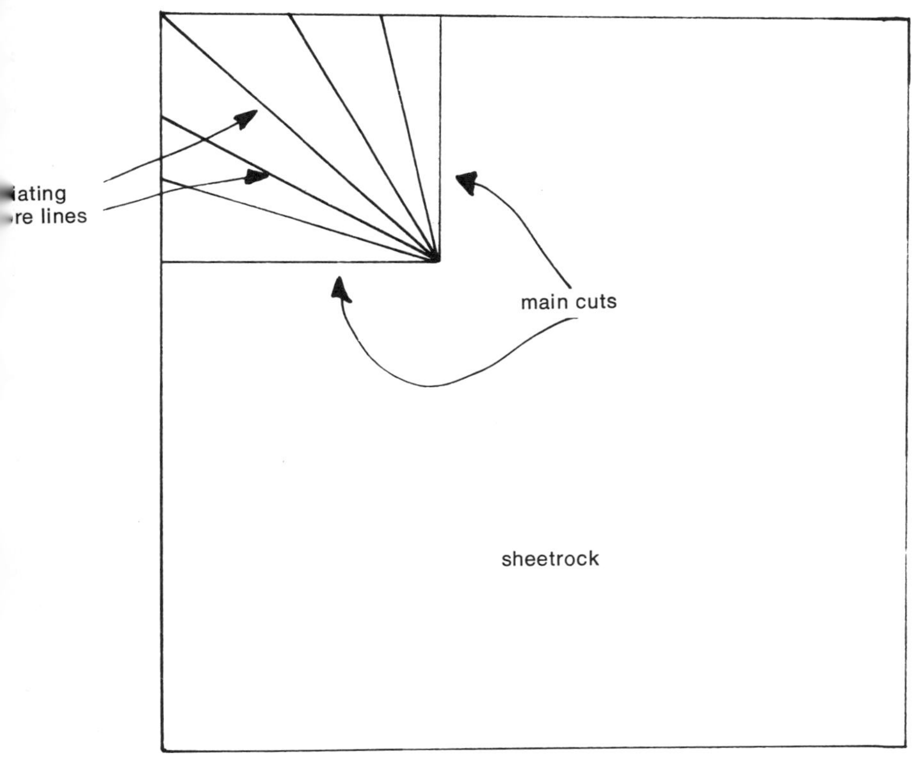

Figure 11.

To remove a section of sheetrock from the whole piece, cut main lines first, then score radiating lines from the corner and break off.

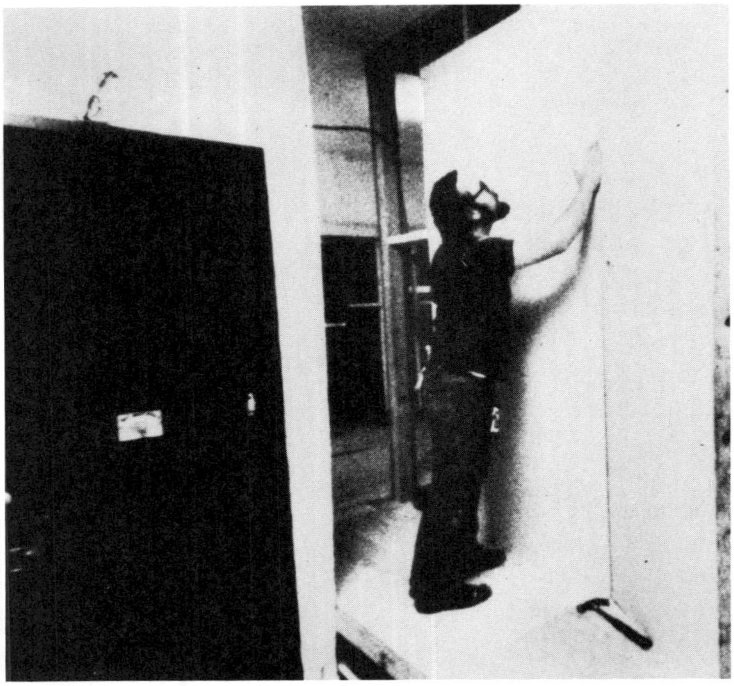

Laying up a piece of sheet-rock that needs hole cut for outlet box. (a) Make sure piece lines up with floor and studs and locate position of outlet box.

(b) Mark and cut hole carefully with knife blade.

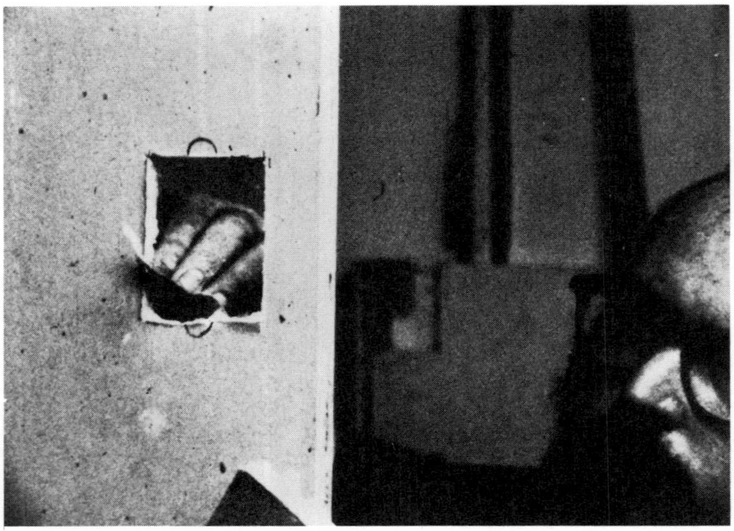

(c) Remove piece.

(d) Install sheet in position and trim outlet opening if necessary.

(e) Nail sheet into place.

cut your main lines and then a series of radiating score lines from the corner; then very gently break out small pieces until you can get an open area within the cutout to break off the remaining pieces cleanly.

You may find it easier to have some help while sheetrocking, especially when you are working on a ladder or directly overhead. Even ⅜″ sheetrock can get very heavy after a while, and working alone on this phase—carrying whole sheets, cutting them, and fitting them into place—can be exhausting. Space the sheetrock nails or screws 8 to 12 inches apart, making sure the heads are slightly recessed so they won't be seen after plastering. As an aid to nailing or screwing sheetrock into place, extend the center-line marks for your studs out onto the floor, so you can sight up from the lines and hit the studs on center when you nail in blind.

Once you've got one side of the wall sheetrocked, you are ready to install any insulation material. For average noise and heating purposes, 3″ or 4″ fiberglass insulation is sufficient and requires a minimal investment. It is available in rolls; simply cut to length as needed and staple in between the studs. For heavy-duty soundproofing, even thick glass wool will not be enough, and you will have to invest in a more elaborate solution. If the esthetics of it don't bother you, you can spray the ceiling and walls with a special acoustical foam of the type used in recording studios. The next most efficient method is to build a double wall—that is, two separate walls built parallel to each other with at least an inch of dead air space and *no* physical connection between them. If there is any physical join between the two walls, sound will be transferred from one surface to the other via the connecting piece, whatever material that might be. This approach, though considered to be quite effective, can be expensive to execute because of the extra labor and materials.

Once your insulation is in, you can close up the walls by finishing the other side of the sheetrock. In practice, you will most certainly build all your walls at once—do all the studding and framing first (for walls, platforms, and the like), then all the electrical and insulation work, then all your sheetrocking.

Framing out complete rooms is really an extension of the simple process we have just discussed. The same techniques apply. To join two walls at the corner, nail the vertical studs together where they meet at the outside (see figure 12). Add an extra stud inside for a corner nailing edge, as well as to tighten up the bond. You can get a nice finished edge on corners by using a special piece of perforated metal called metallic corner bead, or simply metal edge. This substance covers any irregularities in the mating of the two edges of sheetrock and gives you a solid edge to plaster smooth against.

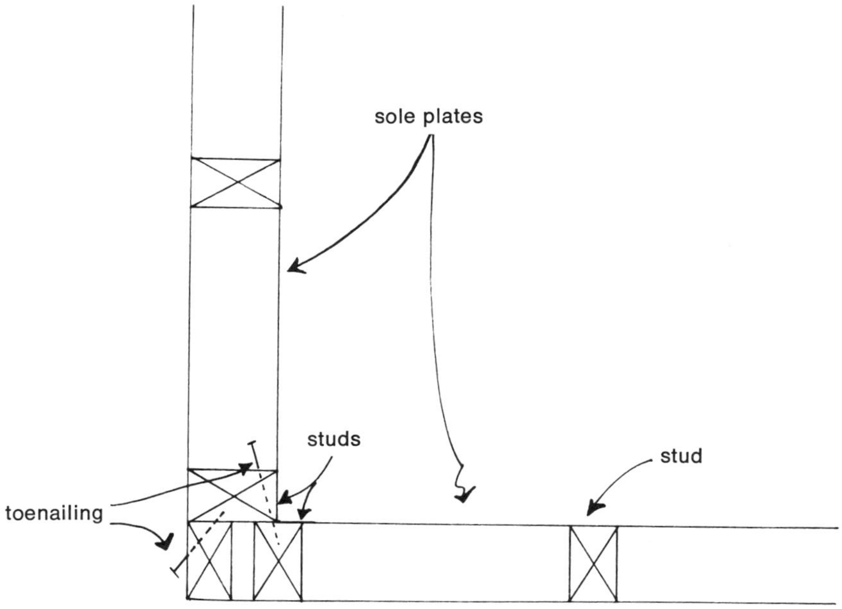

Figure 12.

Overhead view of corner framing joint detail.

.12.
Platforms, Levels, and Loft Beds

While you are doing your partitioning, you will also want to construct any raised platforms, new floor levels, or loft beds. Even if you don't have tremendously high ceilings, you can give your loft a new dimension by changing the floor level in a particular place for a relief for the eye or for making an area separate without actually putting up a wall. A relatively small area can be raised without too much work. The platform can be built with ¾″ plywood flooring, using small sections of ply set on edge for support (see figure 13). Glue and nail the wood together (using spiral flooring nails), let the glue dry, and you are ready for any finishing you might want to do. Some people cover over platforms with carpeting. You can also cover the surfaces with deck and porch enamel, in which case you will need to sand the surfaces lightly before applying the paint. Since raw wood soaks up paint quite readily, you'll have to put down several coats to get a smooth, finished look. Two coats of deck and porch enamel are a minimum; three or four coats would be super! Light sanding in between each coat will contribute to a more finished look.

For large areas of floor raising and level changing, you will have to get into some of the more advanced techniques of housebuilding. These are not terribly difficult, but the cost of the materials is a factor to take into account when planning such construction. The techniques of building new floors are similar to those used in the construction of country houses. First, lay down a row of cement blocks (or several rows, depending on the height of your new level). Then set lengths of threaded rod or carriage bolts into the holes in the cement blocks with wet cement. The bolts should be set in at a depth of at least half the thickness of the cement block and should protrude sufficiently to allow for the thickness of the sole plate plus ½″ to 1″ to accommodate the washer and nut. Once the cement sets, the bolts will be anchored in the blocks. Now place a sole plate over the protruding bolts, mark and drill holes, and set the plate down over the bolts, using large washers to spread the load on the wood (see figure 14).

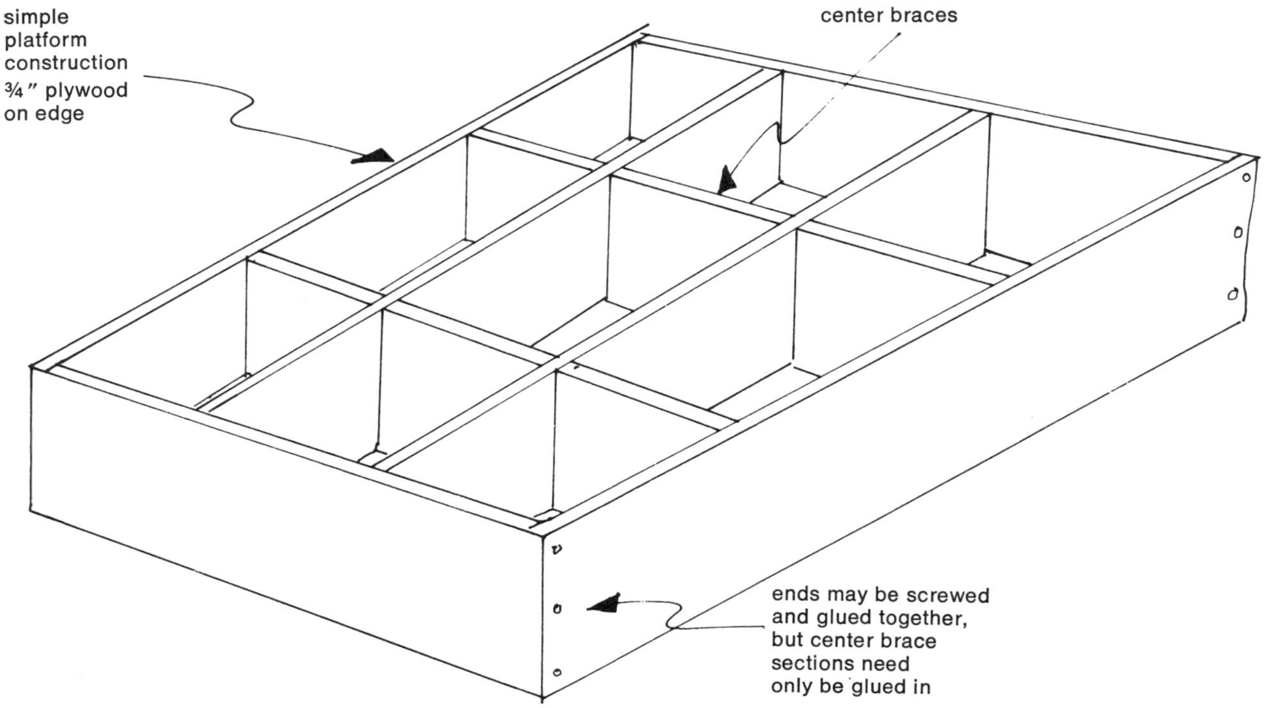

Figure 13.

A simple platform can be constructed from ¾″ plywood set on edge. If you want the platform to support heavy loads, 2″ x 6″s or larger should be used.

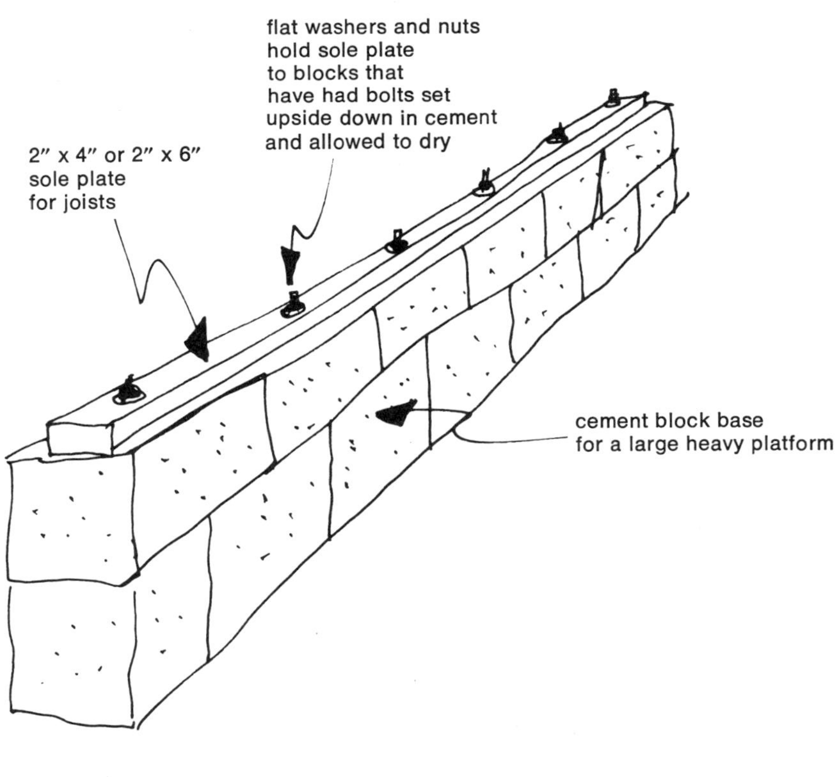

Figure 14.

Raising a large area of floor requires a base of cement blocks upon which sole plates, joists, and new flooring are laid.

Once the sole plate is bolted down securely, you can install your flooring joists. The size and thickness of the joists will depend on the width of the span you are planning to cover, as well as your intended usage for the space. If you are only going to sleep or live over the space, you won't need terribly massive supports—you can use 2″ × 6″s or 2″ × 8″s over a span of 6 to 10 feet. For longer spans, you will have to use heavier beams—at least 2″ × 10″s or even 3″ × 12″s. Also, since these large sizes of wood are not usually kiln-dried, you will have to install either wood or metal bridging to prevent warping, twisting of the beams, and potential loss of underfloor support in extreme cases. Toenail the beams down to the plates, spacing them 12, 16, or 24 inches apart, depending on your needs and design, to allow for laying down 4′ × 8′ sheets of subflooring material, which can be either ¾″ plywood or composition board (Novoply). (Composition board, a type of panel made from compressed sawdust, can be utilized for most of the same functions as sheet plywood. It is cheaper than plywood but provides less support over wide spans. Composition board is also more difficult to cut—you need a very sharp saw blade to make a clean cut without smoking and burning out the saw.)

Once your beams are securely in, you can nail down the subfloor

Cement blocks are needed to support the weight and mass of this large platform.

with spiral flooring nails. If you are planning to conceal any wiring or plumbing under your new levels, you will have to complete it all before you lay down any subflooring, unless you allow a sufficient space—24 inches or more—for crawling under later.

If your ceiling height will allow it and the size of your budget permits, you can even put in an additional floor within your space. This involves bridging the width of your span, wall to wall, with flooring beams and either setting them into the wall on each side or supporting them with vertical beams (see figure 15). If you have exposed brick walls on each side, you can chisel out a hole large enough to accept floor beams. Vertical beams can be installed following the same procedure for a normal floor installation—that is, spacing the supports every 16 inches or, for extra heavy floor loads, every 12 inches.

For new floor levels, you are going to need some way of getting to the new area—stairs or a vertical ladder. The choice depends on individual tastes as well as the amount of space you want to give over to the structure. The preferred angle for a staircase is 30 to 35 degrees—anything steeper may become uncomfortable to climb. If you want an angle sharper than 50 degrees, you should build a ladder, the preferred angle of which is 75 degrees.

For stairs, the maximum recommended riser height (the space between two steps) is 7¾″ and the minimum is 6″. You can vary these figures according to your situation and tastes, but bear in mind that they are standard architectural formulas based on what people have found to be the most comfortable proportions.

Figure 15.

Vertical beams can be used to support new floor level within an existing floor, if ceiling height is sufficient.

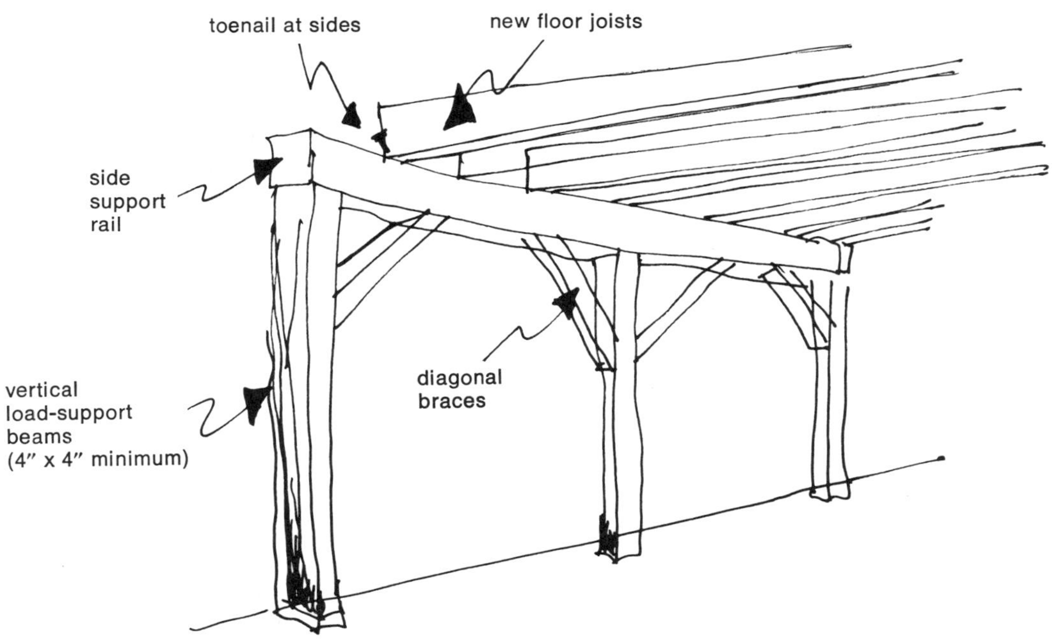

Stairway under construction. Note slots in side rails for steps.

There are *many* different ways to construct steps. The following is one simple method. First, equally divide the amount of space you are planning to bridge by the number of steps you want to have. If you are going to construct steps up to a low platform with a three-step unit, with or without risers, you'll need three equal lengths of tread material—this can be obtained at any lumberyard and is usually either yellow pine or oak with a rounded front edge. (You can also use plywood.) If you are going to fill in the space between the steps with risers, you'll need four lengths of riser material—any scraps of wood you have lying around can be used. You probably wouldn't want to use tread material for this, as it is much more expensive. You'll also need two sheets of ¾″ plywood (or four pieces of ½″ plywood doubled up) cut out in the shape of your steps for side support pieces, to which you will nail your treads. (If your steps are going to be more than 24 inches wide, they should be supported in the center as well.) To connect the unit, you'll first have to attach and brace your supports to the platform as well as the floor, making sure they are vertical and level. Next, nail down your treads, using spiral flooring nails. (You may want to predrill the holes in the steps to avoid splitting the treads when you drive the nails in.) Finally, glue and nail the risers in place.

Unless you don't mind a very simple effect, stairs should be built by a skilled carpenter, who could probably advise you on the various types of stairs that would fit into your scheme. If you are adept at carpentry but not completely familiar with stair construction, you should consult a copy of *Architectural Graphic Standards** for a full description of riser-tread proportions, formulas, and stair data.

If your new level is a balcony, you should give some thought to safety rails and banisters, especially if there will be any young children about the house who might be tempted to jump or who could fall from these new perches.

Even if you don't want to get involved in whole new floor levels, you should consider the option of the ever-popular loft bed, which can save a tremendous amount of space as well as provide an interesting level change for the eye. Loft beds are now in widespread use in regular apartments, especially in older buildings with tall ceilings. The variety and design of loft beds are as different as the individual spaces in which they are built. A quick perusal of the display ads in *The Village Voice* will show you some of the many designs available from carpentry shops and woodworking outfits. It is possible to buy precut kits for easy assembly by the do-it-yourselfer, or you can get a carpenter to come in and build one to your personal specifications.

If you're into doing it yourself, loft beds are simple enough to construct. Styles range from the most simple platform built of 2″ × 4″s and ¾″ plywood with an old ladder for access to more elaborate devices with full-scale staircases, tongue-and-groove hardwood floors, and built-in TVs and stereos. It's up to you and your imagination, as well as your skills and budget, as to how elaborate you want to get.

The most important factor in the design of a loft bed is the physical condition of the location. If it's possible to anchor your supports to the walls, you won't need any vertical support beams. If you can't get a good anchor on the existing walls, you'll have to have posts or columns supporting the bed and rely on the walls only for locating the structure—that is, for preventing any swaying or movement of the bed. In addition to your floors and walls, you have the ceiling; if you can locate a beam, or have exposed beams, you can support all or part of the loft bed there.

Platforms are not always constructed for esthetic purposes only; they may be necessary if you are planning to have a bathroom or kitchen far away from existing sources of water and drainage. Sometimes moving bathrooms or kitchens around is the only possible way to create the kind of space that works best for you. If you want to do this, you may have to have your drainage point fairly high off the ground, thus necessitating a platformed area for that bathroom or kitchen. When boxing in pipes and plumbing for platformed areas, try to think ahead to that time in the future when you may have to do repair work or

* This book is an architect's bible and has a wide range of information on step making and other subjects pertaining to all phases of construction and architecture.

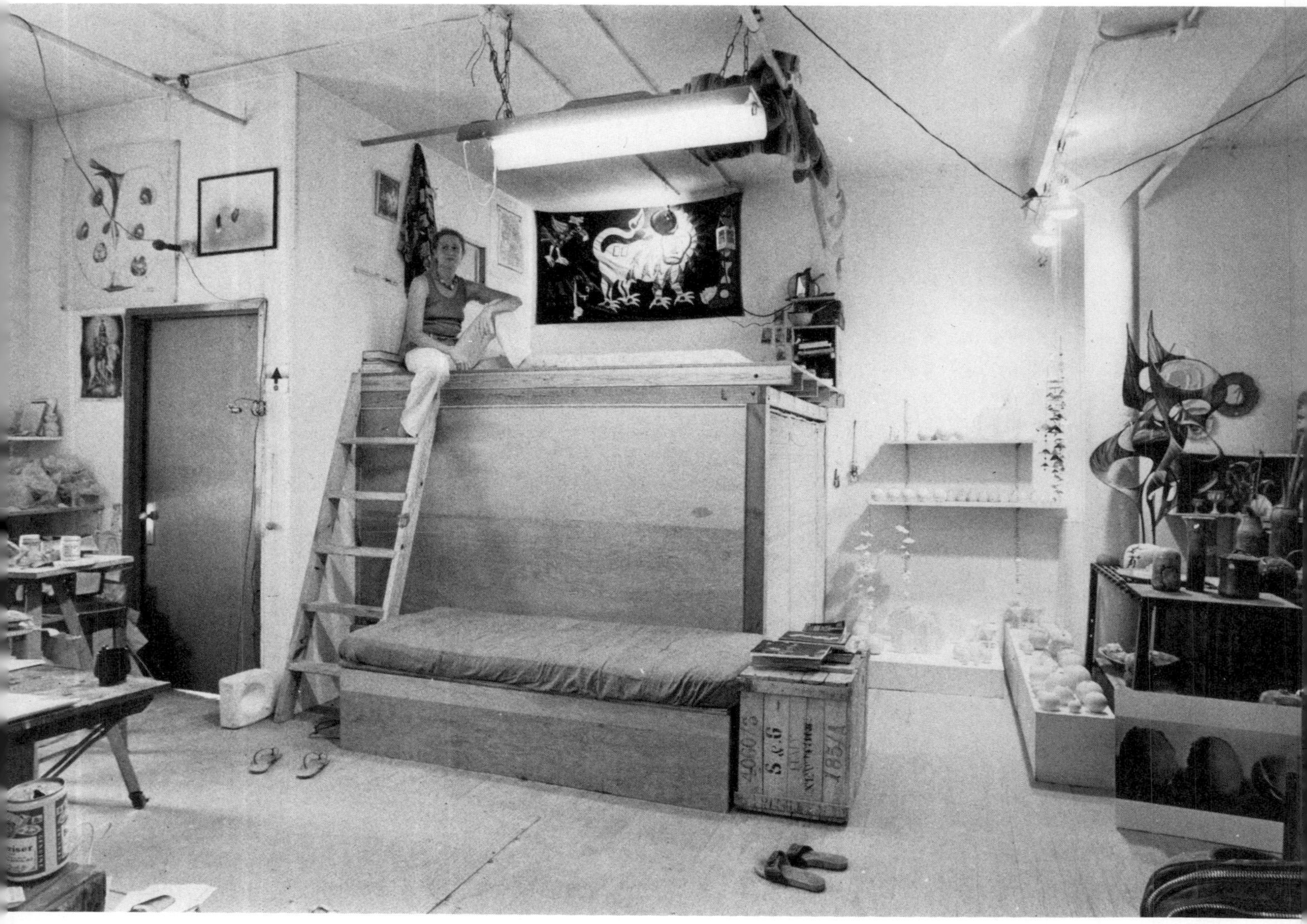

A loft bed combined with storage space and day couch in working potter's studio.

modifications to your system. Try to allow a 24″ crawl space for access to the pipes. A 16″ crawl space will be very tight, and anything less will require an access hatch or removable floor section. In laying out plumbing, try to provide some cleanout plugs for later access. Drains do clog up after long periods of use, especially in showers or baths and in kitchens. With a little forethought, you can avoid having to demolish a floor or wall later to get at a pipe for a repair.

A loft bed that uses the walls of the corner of the room and ceiling beam for support. Note simple vertical ladder.

.13.
Completing Walls and Electrical Work

When all the electrical and plumbing work within your walls and partitions is finished and you have installed the second side of sheetrock, you are ready for the taping and plastering of your new walls. This task is a vital one if you want to avoid any separation between the boards of sheetrock. There is a fair amount of skill involved in getting joints with no irregularities, but most people can gain the necessary skill while actually doing the work. A premixed joint compound will make the job easier, as it stays workable much longer than conventional plaster. Stir the stuff and add a little more water to get a smooth consistency that will be easy to lay up. Professionals lay up the first coat of compound very thinly with tape paper, let that dry, and then lay up an additional coat or two of joint compound to get a smooth finish. It will probably take the less skillful loft dweller a fair amount of time to get an acceptable finish. Try not to put up too much compound at one time; a series of thin coats will look better than one heavy coat, which may require lots of sanding to take down any bumps or irregularities.

Begin by laying up a thin layer of the joint compound. Then, starting from the top of the wall seam, place the paper joint tape over the wet compound and press it down, using a 5″ tape joint knife; work from top to bottom, trying to maintain as smooth a finish as possible. Go back over the tape, laying back down any compound that has accumulated on the knife. You'll get a nicer finish if you keep the knife at a fairly flat angle to the board (30 to 45 degrees). For your second and third coats, you may want to use a 9″ knife.

For joining inside corners or walls to ceilings, fold the paper tape along its imprinted crease to make a right angle. Then apply compound on each surface and lay up a piece of folded paper. Run the knife down both sides and repeat, trying to leave a little of the compound to cover the tape for a smooth, perfect corner. A little time spent on this will be worth it.

To get smooth walls, two or three coats of joint compound will be needed.

You can hand-sand between coats, but if you are going to be doing a lot of sanding, you'll find a cheap oscillating sander will make the job go faster. A word of caution here. Wear a dust mask and goggles when you sand down joint compound, as its high asbestos content is potentially harmful. Professional tapers have considerably shorter life spans as a result of not wearing proper protection for their lungs and eyes. Some tapers now do as much smoothing as possible with a big sponge while the compound is still wet, in order to cut down sanding later on. Also, when you are sanding, be sure not to sand away the paper covering the sheetrock—if you do, after you are through painting the walls the surface will appear much rougher than the areas that have the paper covering intact. At this time you should also plaster over any depressions made by nails. Don't try to plaster the holes smooth the first time, as the compound shrinks when it dries. You should hit these spots each time you do another coat on the seams.

Once all your taping and plastering is finished, you can install and wire up the electrical receptacles and switches in the new walls. Don't put any cover plates on yet, unless you want to paint them the same color as the walls. This may be desirable in living room areas, but in bathrooms and kitchens you may want to have chrome cover plates, which are best installed after you are finished painting.

If you haven't done it already, you should now finish off any other electrical work on existing walls and ceilings, such as new wall outlets or circuits for overhead lighting. Any new surface-mounted electricity will probably run inside EMT (electro-metallic) thin-wall tubing, which is rigid and must be bent by means of a thin-wall tubing bender (or "hickey" bender) to conform to whatever surface it is being attached to. This is readily done following the instructions included with the tool. Flexible cables, such as BX or Romex, are usually acceptable only for use *within* walls, not for surface mounting. If in doubt, check your local electrical codes first or consult an electrician.

Working with EMT is not difficult, and it looks fine when finished and painted over neatly. Start work at the supply box (fuse or breaker box) and progress from there. At each point where you want an outlet, you'll have to install a junction box—a square or rectangular metal box with a series of knockouts prepunched for acceptance of the connectors necessary to mate up with the tubing. Fasten all boxes and tubing securely to the walls, ceilings, or baseboards, using screws with lead shields for brick or masonry or toggle bolts for hollow walls. The size of the tubing will depend on the number of wires and circuits you are planning to run. The most common size used for average work is ½″, which can accommodate up to four No. 12 wires or two or three No. 10s. The ¾″ size can be used for almost everything else, except perhaps your supply line from the basement, which may require 1″ or even 1¼″ tubing. Unless you have on-the-job experience as an electrician's helper, you should not attempt to put in your own basement-to-floor supply.

Once the tubing and boxes are installed, you're ready to pull the wires through the tubing. For this you'll need to feed a snake—a length of thin spring steel—through the tubing until it comes out at a junction point. Attach the wires to the end of the snake by bending the ends over and taping them together with plastic electrician's tape. Have a second person help you either pull the snake or feed the wires into the tubing. This will lessen the possibility of stripping off insulation and thus prevent any shorting out of bare wires against the metal box or tubing later on. Use a liberal amount of pulling compound (available in plastic squeeze bottles) on the bundle of wires as you push them into the tubing from the box. The compound will considerably reduce the friction of the wires being pulled inside the tubing (especially if there are bends in the sections you are pulling through) and will make your work go easier. Once you've pulled all your wires, you can wire up the various outlets and switches and then install cover plates.

A note on installing ceiling lighting tracks. If you are surface-mounting them, you should put them up first and then connect the tubing and switches necessary for hooking them up to their sources of power.

.14.
Finishing Plumbing Work and Toilet Installations

With some of the walls finished and your electricity, basic drainage, and water supplies in, you're ready to put in the fixtures and start washing again! Hopefully, you've located the hot- and cold-water supplies and drainage points in their proper places and are now ready for hooking up your fixtures.

Sinks are pretty easy to install these days, with one note of caution. If you are planning to use a stainless-steel sink in a 1½″ butcherblock counter, you will find that the screws for the mounting clips provided with the sink will not be long enough. They are designed for use only with ¾″ countertops. The screws are special reverse-direction threaded types, not available in any hardware store. The manufacturer of the sink may make longer ones, but the time spent trying to communicate with a factory in Wisconsin may be too much to contend with. You have two choices. You can abandon the idea of a stainless-steel sink and go for a cast-iron enamel-covered unit that is installed with a line of adhesive rather than anchoring clips. Or, if you want the stainless steel, you can drill a hole in the edge of the slab of butcherblock at a depth of ¾″ to allow the clip to be inserted and tightened, and thus utilize the screws provided.

Since almost all sinks today are drop-ins, you'll have to cut a hole in the countertop you are using. Many sinks come with an exact template —a paper pattern for you to lay down and mark a cutting line with. If you are using plywood for the counter, you can cut rounded shapes with a saber saw and make straight cuts with a circular saw. If you are using a butcherblock counter, the sink cutout can be a problem. Maple is an extremely dense wood, and you will burn out a cheap saw trying to cut a slab of it. Unless you have a heavy-duty saw with a very sharp blade, you should not try to do this job. Stores that sell butcherblock will usually cut the opening for a slight fee (but quite worth the savings) if you bring them the template or the sink itself.

With the sink in place, hook up the hot and cold water and drainage.

Finished bathroom with new toilet and sink installed.

Most handsets (faucet and mixing spout assembly) mount to the sink itself with a rubber flange on top and are drawn tight with two flange nuts and washers from below. You can cheat on this by installing the handset ahead of time, before the sink is in position, and thus the nuts are more easily accessible. Otherwise you will have to use a basin wrench and work from below, and you'll need a helper to hold the handset in position while you tighten the retaining nuts. You should own a basin wrench anyway, because you'll need it to connect the water supplies, which in most cases will be flexible tubing with compression fittings. Known often as "speedy connectors," these chrome-plated supplies have a flanged piece on one end that is screwed up under the sink to the base of the handset inlet. The other end is straight tubing that should be cut cleanly with a tubing cutter so the compression ring can slip easily over the end.

The compression ring comes with a supply valve called an "angle stop," which provides a convenient way of making the transition from the supply line to your sink. With the valve in place, it is possible to get your water back on without having to have the sink fully installed. The flexible tubing means you don't have to figure out to the last inch exactly where the sink is going to be. The closer you are to your supplies, the less additional material you'll need to complete your hookup. Flexible connectors are commonly sold in 12″, 16″, 24″, and 30″ lengths.

To connect the drainage on a bathroom sink, you'll need a tailpiece with or without a sink-stop lever. This is usually a two-part assembly, with one piece being screwed into the other from the inside of the sink. Place a little ring of plumber's putty under the edge of the flange that screws down from the top; as you tighten the flange, the putty will be squashed flat and the excess squeezed out, thus providing a nice seal between the porcelain sink and the metal drainage pieces.

For a kitchen sink you'll need a strainer assembly, which is installed with a little roll of putty under the top flange edge. Have a helper hold the basket in position with a pair of pliers while you screw on a large flange nut from below. A pair of wide-open Channelocks will do the job very well, although there are special wrenches for this job.

To connect the strainers to the trap, you can use either PVC fittings with slip joints or rough-brass slip fittings. With the correct lengths, elbows, and offsets available, you can make up the necessary distance to the trap. Just remember to maintain proper pitch (¼″ for every foot) for good drainage, and keep the number of joints and connecting pieces to a minimum. Slip joints can come apart and can be easily dislocated unless they are properly strapped up and protected.

Even if you already have a toilet installed in your loft, you may want to replace it with a more modern unit or move it to a new location. Despite their seeming complexity, toilets are actually quite simple to install, and setting up drainage for a new or changed location is not

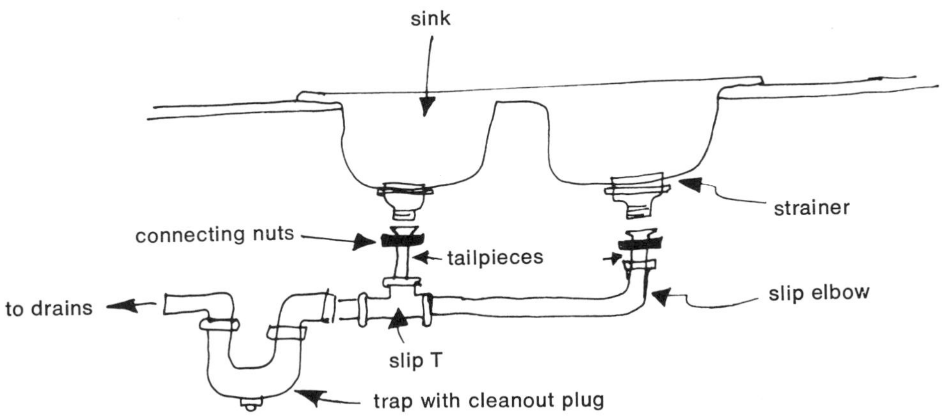

Figure 16.

For a double-basin kitchen sink, connect a tailpiece to each of the strainers. Then connect a slip elbow from the end of one tailpiece to a slip T under the other tailpiece. The two sinks are now connected to one another drainage-wise. (Make sure to maintain proper pitch from one sink to the other.) Make up the necessary distance from the end of the T to wherever the trap is, using either brass slip fittings or PVC pieces.

difficult. The same principles apply to toilet drainage as to sinks and showers; the difference is mainly one of size. Use 3″ pipe minimum for short runs and 4″ or 5″ pipe for long runs or runs with bends. Pitch is the same—even a little less pitch than for sinks (say, ⅛″ per foot of rise) is considered better. Just watch for too many bends, which might become points of restriction later on.

Unless you feel pretty proficient in drainage work and the use of PVC pipe, you should not attempt this sort of work yourself; if you can find a craftsman to do it and let you act as helper, you might be better off. For the nonprofessional plumber, PVC is the only easy way to do toilet installations away from the source of the main riser. Normally, this sort of work would have been done in conventional cast-iron pipe with oakum and hot-lead joints, but as mentioned earlier PVC has simplified all that.

If you can get access to the point in the main riser where the lead bend fitting for the previous toilet went into the stack, you can remove the old fitting and start fresh with a PVC-to-cast-iron hub adapter. The old lead fitting can be removed more easily if you use a butane or acetylene torch to melt out the hot lead. While applying heat, grip the sawed-off edge of the fitting with a pair of heavy pliers and twist hard. If it is really in solid, try chiseling away some of the material between the two fittings. It will take some doing, but you should be able to get it out after a while. Once you do, clean out the hub part of the cast iron with a wire brush and rags. In the meantime, be sure no one above is using the drain while you have it open, or you may be doused with a load of unmentionable matter should someone above you flush. With the cast-iron hub clean and dry, insert the proper-sized PVC adapter with a few strands of oakum strapped around the end. Using a small chisel (there are special chisels for this, but almost anything will do, including a screwdriver), tamp the oakum down all around the fitting, trying to be as even as possible (see figure 17). Keep feeding the oakum in, packing it down until the fitting starts to feel tight and you are within about half an inch of the surface fitting. Now, on top of

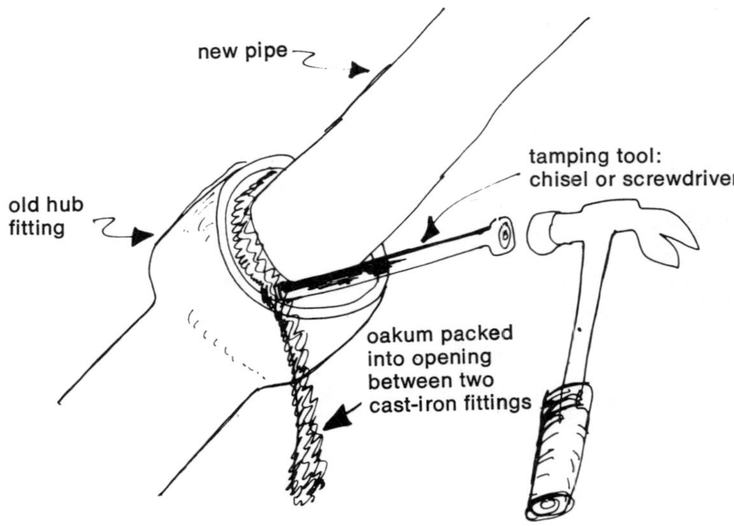

Figure 17.

Oakum is packed in the space between two fittings to make a water-tight seal when used with poured molten lead or shredded lead wire.

the oakum, lay in shredded lead wire, again tamping and driving it into the opening as evenly as possible.

Now you can start laying up the rest of your drainage piece by piece. Once you've got your first few sections secured fairly well, you should go back to the PVC hub fitting and repack the shredded lead—you'll find that small movements during assembly may have loosened some of the packing. Retamping this joint and applying a good seal of A&B plumber's epoxy will ensure trouble-free drainage for years to come.

If the new drain is strictly for the toilet, continue on to the point where your new installation is to be. Check the rough-in of your new toilet—that is, the minimum distance from nearest wall to the center line of the toilet flange. For some units, it will be 12 inches; for others, 13 or only 11 inches. Offset floor flanges are usually available only in bronze, for lead bends. Since this drainage system is probably going to be under a platform of some type, you will have to put down the piping first, with an elbow directly under the center line of your new toilet rough-in, and install a vertical section of pipe that can stick up above the new floor line. Once the new floor has been laid down, the vertical pipe can be cut off flush with the floor and the toilet flange screwed down to the floor. The flange provides something in which to mount the toilet bolts that fasten the bowl down. Next, lay down a fresh wax ring on the porcelain flange of the toilet itself and place it over the floor flange with the two bolts protruding through the holes in the base of the toilet (see figure 18). Gripping the bowl firmly, twist gently from side to side, bearing in mind that you are seating the wax ring and making a waterproof seal between the bowl and the drain itself. Install and tighten the nuts *evenly*, a few turns per side at a time. In addition, when you install a toilet, you should set the bowl down in a

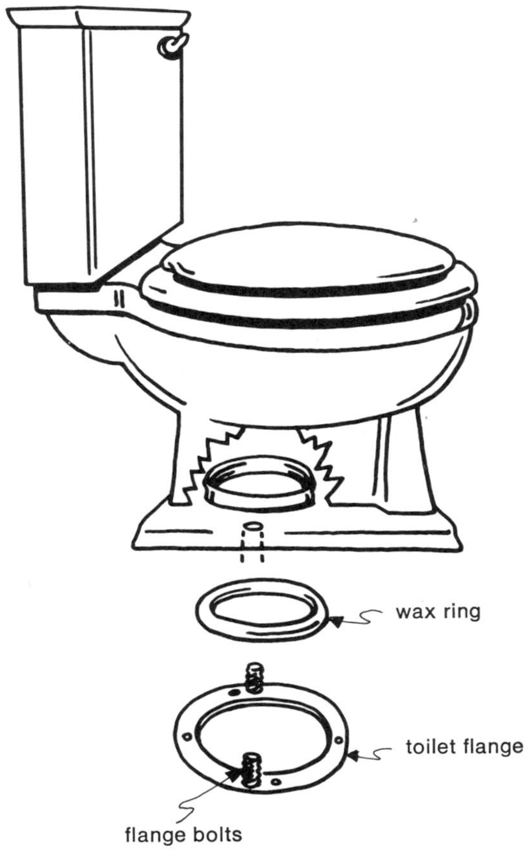

Figure 18.
Toilet installation.

very thin, liquid bed of plaster of Paris, as an aid to distributing the load on the vitreous-china edge. With the bowl down, hook up the cold-water supply at the base of the tank. This can be readily done with a ⅜″ compression angle-stop valve and a short length of flexible toilet-connector tubing. Turn the valve on and check for proper flushing action. The ball will probably have to be adjusted to get the correct water level—too little and you may not have good flushing action; too much and you'll have a constant overflow sound as excess water trickles over the tip of the fill tube. Check for any leaks at the base; if there is a leak, it is the result of an imperfect seal at the wax ring—you may need a double ring or an extra deep wax ring. The good thing about a PVC system is that it is virtually maintenance-free; unlike cast iron, there is no metal to rust and corrode away—just an impervious plastic that will last almost forever.

.15. Painting and Finishing

Most people think that the painting and finishing of their loft will be a quick and easy final stage, but it can take a surprising amount of time. The great thing about the painting phase is that anyone can do it, thanks to rollers and easy-to-use latex paints. There are a few things you should know about paints before you begin. First of all, latex (water-base) paints may be easy to apply, but they cannot be used on certain surfaces—namely, walls or ceilings that have been discolored by staining or seepage. Applying water-base paint over stained surfaces will not work, no matter how many coats you put down. You must use an oil-base paint first to act as a seal against the stain. (For extreme cases there is a special alcohol-base stain-sealant paint.) Once this is done, you can go over the oil-base paint with a coat of water-base paint.

Try to buy all your paint at once in large five-gallon tubs, and see if you can get a break on the price—buying paint by the single gallon at most hardware stores is the expensive way to go. Many paint factory outlets, building supply houses, and similar places offer breaks on large orders; if not, they will usually throw in such things as mixing paddles, buckets, and painter's hats free of charge.

When painting, work from the ceiling downward; that way you pick up any drips as you go along. Most open areas can be done with a roller on an extension pole. Ceilings can be done this way too. If you have an embossed metallic ceiling, or one with lots of pipes and tubes, you can use a deep-pile roller and keep going over it from different angles. Corners are best done by brush, then blended in by roller.

Attention to detail during painting will make a big difference in the overall look of your place. Careful use of masking tape to separate different areas of color is really worth the time it takes. If you are planning to paint the ceiling a darker color than the walls, you should mask the line and hang drop cloths with tape.

In addition to painting walls in various shades of color, you can approach walls as canvases and apply "artwork" directly to the wall surfaces in the form of murals or "supergraphics."

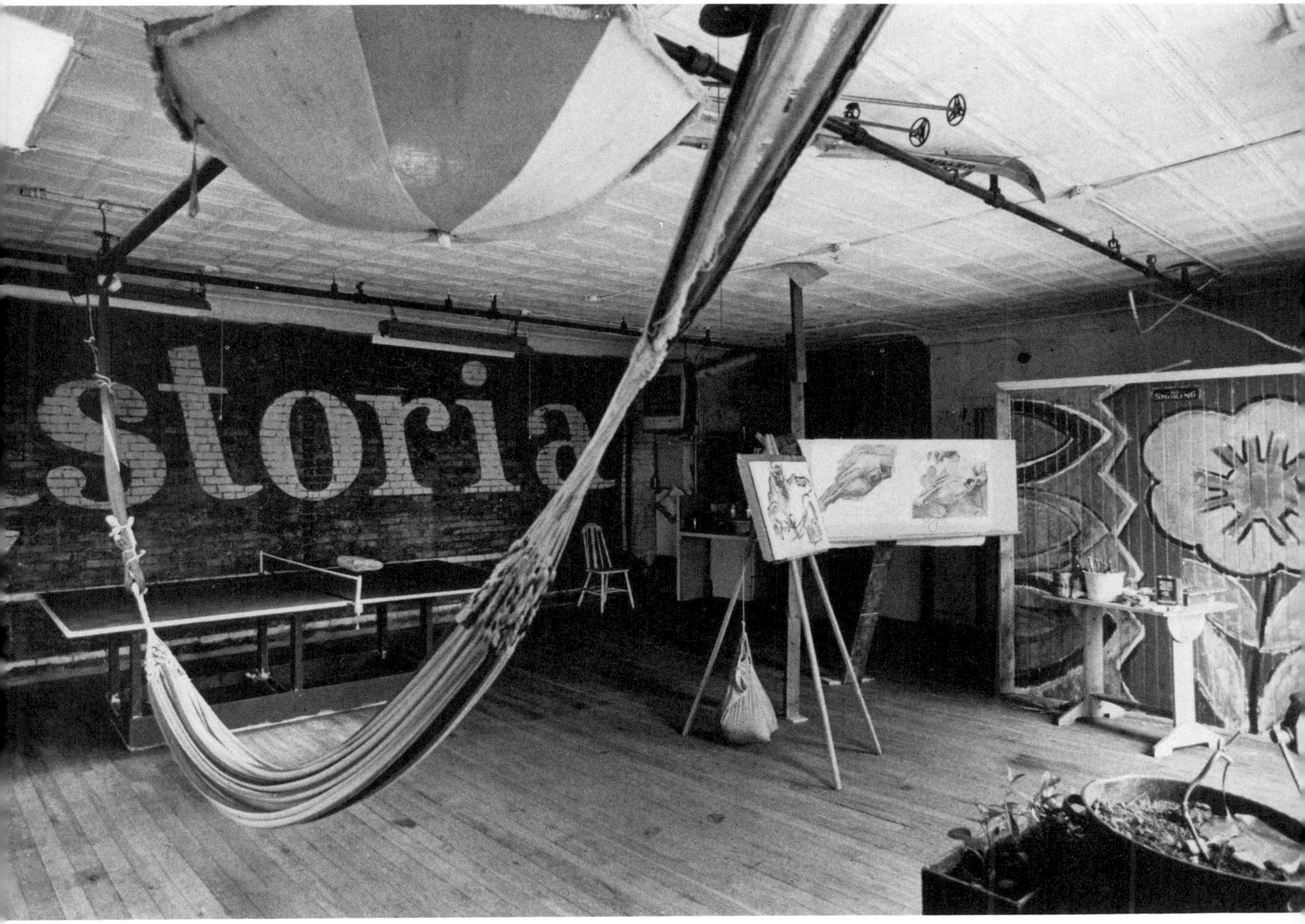

This loft dweller uncovered an original building advertisement when he stripped the plaster off the wall. Another wall was used as a canvas for a mural.

Window frames should be masked off if you are planning to paint them a different color than the walls. Make sure all loose paint is removed and the frames are scraped clean before you begin; if it will fit into your plans, use a semigloss or full-gloss oil-base paint, which will stand up better around a window than the water-base type. You should also mask off the glass areas if you think you are going to splash a lot of paint about; but if you use a special sash brush and have a steady hand, you may not need to mask off the glass. Masking tape can get glued on to glass and may have to be scraped off with single-edge razor blades. You'll probably need the blades even if you don't mask off the glass, but you may find it easier to scrape off a few dabs of paint than a few hundred inches of tape.

Before you paint, you should be sure that all your surfaces are

Sanded and refinished floors are lovely but costly.

properly prepared. If you have a badly cracked wall with many undulating surfaces and you don't want to invest the bread in new sheetrock, you might think about giving it a texture with the application of either Texolite or ready-mix wallboard (joint) compound. Wallboard compound is excellent because it dries slowly, thus enabling you to do a large area at one time. The compound can create a nice effect on an otherwise ugly wall, which can then be painted over easily, and will serve to bind the wall together. (Incidentally, Spackle is very expensive —about $12 per gallon—and doesn't go too far. You'll be better off using ready-mix compound, which will do the same job and runs only about $1.60 per gallon when bought in five-gallon tubs. Joint compound is a good all-purpose materal for patching and plastering—you'll probably be using a fair amount of it anyway, and the empty five-gallon buckets make great tool carriers once they're all cleaned out!)

A few coats of deck enamel give your floor a finished look inexpensively.

Another interesting way of handling unsightly walls is to cover them entirely with fabric. To do this, measure the area very accurately. You will not be able to get fabric wide enough to cover an average-sized wall. Therefore you will have to sew together equal lengths to make up the width. Then simply staple the fabric to the top and sides of the wall. You can trim the top and bottom or just fold under the edges. Remember to pick a fabric that will look OK seamed together.

After you're finished painting the walls and ceiling, you're ready to put the last touches on your loft. This includes painting the floors if you haven't already sanded and finished them. Now is the time to install electrical cover plates and to hang any doors that may have been removed for finishing or trimming. Allow for any door sill that you may want to install later on. Door-frame moldings can be cut, stained or painted, and then installed. Put in your locksets and knob mechanisms too.

.16. Special Effects with Custom-Made Pieces, Lighting, and Plant Environments

The term *custom work* or *custom pieces* connotes something out of the ordinary, something unique and handmade—and, for the craftsman, something to be proud of.

For most people, the word *customized* calls to mind custom-made or built-in furniture—couches set into corners; beds built into preexisting niches and irregularities of a wall; kitchens constructed in every size, shape, and description. All these things are great ideas, as they very often make use of preexisting shapes or otherwise unusable space. Part of the loft concept is making use of dead space and converting it to function usefully for you. At the same time, you have that opportunity to express yourself artistically and philosophically in the way in which you carry out your ideas—your choice of materials, finishes, techniques, and so on.

The secret ingredient in customized projects is time—backed up by some funds to obtain necessary raw materials and supplies. You have to have enough time to do it right and enough time to go that one step further in skillfully executing the work. Also, some time spent in careful forethought is better than languishing in the throes of hindsight gained through bitter experience!

CUSTOM BATHTUBS

There are a tremendous number of ways to apply yourself to the creation of customized environments with a loft space. One of the things that appeals to a lot of us in this hedonistic society is a custom bathroom. This is nothing new in the annals of mankind—the Romans started working on bathrooms a long time ago and perfected a lot of technology still used today (tiled baths, for example). For the loft dweller, a custom bathroom or tub is a welcome escape from the confines of modern buildings or old tenement bathrooms. The average

Weeks of work and sanding were required to create the luxurious mahogany counter below.

bathroom today is very cramped, with a minimum of space to perform a variety of functions. In a loft you are at liberty to create a much more spacious and humane environment for bathing and washing.

One of the ways to really enjoy yourself is to have an oversized bathtub, made up in whatever shape you want, to fit into any corner or space you might have. All this is available to any do-it-yourselfer who can cut plywood and put it together in a boxlike shape. Basically, you make the form for your tub in wood, coat the surfaces with fiberglass resin, then tile over that with conventional mastic and thin-set tiles for a custom-built tiled tub. If you don't want to bother with tiling, you can still have a nice custom tub by just laying on several coats of fiberglass to waterproof it and give the tub additional strength.

In addition to the normal tub-filler and shower assembly, you can install special shower heads, the popular whirlpool Jacuzzi heads, or steam machine inlets. Though these last two items are fairly expensive (about $550 for a three-head Jacuzzi assembly and pump you install yourself at the sides of the tub, and about $475 for a steam machine capable of steaming a 6′ × 6′ space), they are rather nice and certainly provide the finishing touch for a far-out bathroom. If you're throwing $10,000 into fixing up your loft, it's really not all that extravagant.

If you are going to build your own tub, you should calculate in advance the size of the tub and the volume of water it will hold to make sure you will be able to supply enough hot water for it. If there is no main hot-water system, you'll probably need a good laundry-sized unit —approximately 50 to 75 gallons—with a quick recovery rate. A 5′ × 5′ or 5′ × 6′ tub is about as large as most people want and can afford to heat, but if space and conditions are agreeable, larger sizes are certainly possible. The weight of the water is an important factor to consider. A cubic foot of water weighs 62.4 pounds. Thirty or forty cubic feet of water add up to a lot of stress on the floor. If you are in a creaky old wooden building, you shouldn't be doing this sort of thing.

Most old lofts, with their industrial floor loadings, will be fine for tub weight support—unless the floor is really sagging a lot, in which case you should be careful. The plywood frame of the tub should be firmly held together with "hoops" of 2″ × 4″s that are screwed into the frame with solid No. 12 screws every 6 to 8 inches (see figure 19). (A power screwdriver or Versamatic attachment for a drill will make the job go much faster and easier.) Be sure to use at least exterior-grade plywood, if not marine-quality plywood, as these types are made with a non-water-soluble glue. The glue in interior-grade ply may dissolve if it gets wet; then the ply will start to delaminate and you'll have some problems.

Try to stress the weight of the tub between two walls if possible, or at least in a corner with an abbreviated triangular front. The less the tub is subject to movement, the less the likelihood of any cracking or leakage later on.

Detail of shower and tub filler assembly.

Figure 19.

Detail of construction of plywood and 2″ x 4″ frame for oversized custom tub.

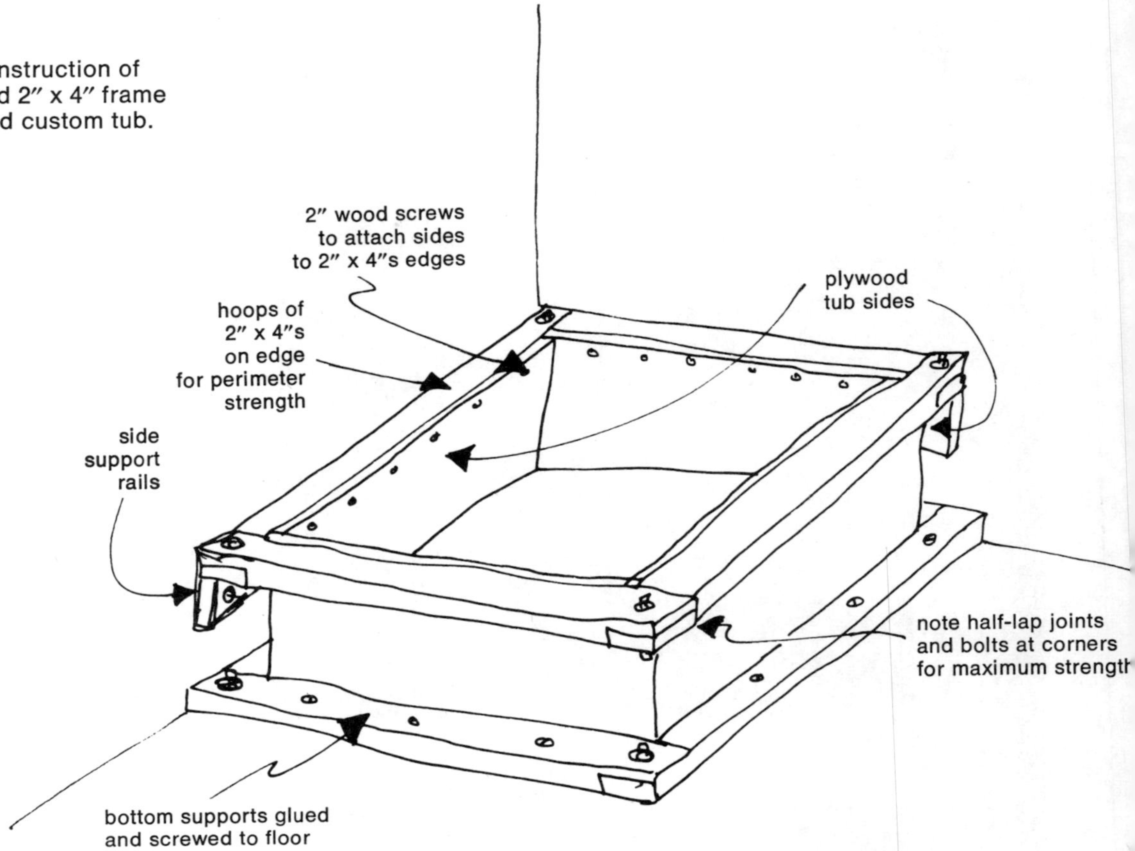

Once you've got the basic tub shell screwed together and stressed properly, you can apply the waterproofing and structurally strengthening fiberglass resin coat. This is the same material used in the construction of boats and cars, and when dry it is immensely strong as well as being waterproof. To waterproof a bathtub, apply the liquid resin with a wide brush and let it dry until it no longer feels sticky to the touch. Air temperature and low humidity is very important to ensure proper curing of the resin. A minimum temperature of 70° F. is required over the period of curing. If you are planning to do such work during the winter, you will have to make some arrangements to keep the work area warm. If you are in a very damp area, it will take longer for the resin to set properly. In addition, the fumes from the resin are quite strong, so you will need to have direct ventilation to the outside air. If you are working in a steam-heated area, you should apply the resin during the day, while the heat is at its strongest level (even apartment houses lower their steam heat at night).

If you've never worked with fiberglass resins before, there are a few things you should know. To begin with, once you've mixed in the catalyst (the stuff that makes it harden), you've usually got only about

Firmly screw the sides of the tub to the 2″ x 4″ retaining hoops. A screwgun makes the job go faster. Below, tub is well braced between two walls and ready for waterproofing.

20 minutes of working time before the mix will turn into a solid block of matter. Mix a small quantity to begin with and see how much you can get done before it hardens. You can vary the hardening time a little by adding more or less hardener. But don't stray too far from the manufacturer's recommendation or the resin may not cure at all; even worse, it may harden too quickly, leaving you with an expensive block of resin! Medium-quality fiberglass resin sells for about $8 per gallon; a more expensive polyester resin sells for about $12 per gallon. As with many materials, there are pros and cons to each type. The fiberglass resin is cheaper and sets and cures quite quickly, but it has a terrible noxious odor that few people can tolerate for very long. You have to lay the stuff down quickly and run out of the room (which hopefully has a door and an open window). The polyester resin has very little noxious odor, and for some people this is a definite plus. However, it takes several days to dry properly and may even have to be helped along with an array of heat lamps.

You can apply the fiberglass or polyester resin with a brush (which will have to be thrown away afterward, as it will harden like a rock). Start with the top edges, then continue down the sides of the tub, followed by the floor area. For extra strength, you can lay folded strips of woven fiberglass reinforcing tape along the corners and bottom edges where the sides of the tub join the floor; it's a little bit of insurance for the future. All these resins have a certain amount of flexibility, as they are used on boats that are subject to more extreme stresses than those that will be encountered in a loft building. If the building settles a little bit, your tub will not split apart from the slight additional stress.

A word here on pitch for your tub. Without proper pitch your tub will not drain completely, and you'll be left with a little pool of water after each use. To avoid this, try to build in some pitch while you construct and lay the tub in. After you've got the tub fiberglassed, if you still don't have enough pitch, you can correct the situation by mixing up another batch of resin; this time add some plaster of Paris, a little at a time, which will thicken the resin without letting it get stiff too quickly. Pour the thickened resin along the line where you want to build up slope and smooth it out using a rubber squeegee or a piece of wood with a clean cut edge—it'll be like smoothing out a cement slab. Incidentally, this mixture may be applied as a final coat for any situation using resin where you have laid down matting or woven roving that has left a rippled or unsmooth finish when dried. A truly smooth surface is highly desirable if you're going to be laying down ceramic tiles in the tub. If there are any bumps or irregularities in the surface, you can grind them down with a belt sander and coarse-grit paper.

Tiling the tub is not absolutely necessary, since the fiberglass or polyester resin will give you a perfectly waterproof and serviceable finish. However, many people like the idea of tiles, and it is a way to get quite creative if you so desire. There are thousands of different types of tiles on the market today, but not all kinds are applicable to tub and shower installations. Tiles come in varying thicknesses; the thinnest are usually the best for tub jobs, unless you are using the conventional mortar-mix installation method, in which case you can use any thickness tile. For our purposes, we're going to stick to the thin-set tiles that can be installed with mastic adhesive. This is really a much simpler method than the old-style wet-mix mortar, and most people will find it quite satisfactory.

There are many different manufacturers of tile in this country; one of the biggest is American Olean, in Olean, N.Y., which has a very good book (actually a thick pamphlet) showing the different tile styles and sizes as well as a complete line of inner and outer corner and edge pieces to give your work a customized look. For the unlimited budget, there are some terrific tiles imported from Europe (mostly Spain, France, and Italy). Very little inventory is carried on most tiles (especially the imported kinds), and you will probably have to order them in advance. Try to calculate fairly accurately the amount you will need; running out of tile is an immense bummer—it usually takes anywhere from several weeks to several months to get tiles shipped, and the color may vary substantially from batch to batch. Get all you need at one time, and don't forget to figure in a percentage for cutting and breakage.

To lay down tile, spread a layer of mastic onto the surface you are tiling, using a notched trowel to create the thin beads of mastic necessary for good adhesion. If you don't use a notched trowel, you will be putting up too much mastic, and it will take too long to dry. In addition, when you lay down your tile, the excess mastic will squeeze up between the tiles, making it impossible for you to apply grout properly. Don't spread too great an area of mastic, as it starts to dry the moment you apply it. Small tiles usually come in sheets, either backed with gauze or held together with beads of glue that have been applied at the factory. These sheets make the application much easier, in terms of both aligning individual tiles and keeping all the tiles straight. Use a rubber mallet to tap the tiles flat and stick them down—this helps on vertical installations, of course, but even for flat surfaces it's not a bad idea, because you want to get the maximum possible adhesion to the bonding surface.

At certain points, no matter how carefully you have measured and laid out the tiles, you will wind up having to make some cuts. There are some big water-fed tile-cutting machines available, but if you are just doing a small amount of work a hand-operated tile-scoring and cutting device will be fine. The cutter has a sharp wheel similar to a glass cutter. To make a cut, hold the tile in place and score a line where you want to break the tile. Then crack along the line. The piece to be retained should be supported underneath the score line, while the piece to be broken off is left in the air; downward pressure on this piece will result, hopefully, in a nice, clean break. If you are not completely successful the first time, do not despair; after a few cuts and breaks you will get the knack. For small cuts, or for cleaning up cuts made on a tile cutter, you can use a pair of nippers, similar to a pair of pliers. Remember that you don't have to make perfect cuts in the tile to allow for the protruding pipes of a shower head or tub spout; there are usually escutcheon rings around the devices to cover an extra inch or so of the wall.

The next step is filling the spaces between the tiles with a grout mixture. Be sure that all the tiles are securely glued in and the mastic has fully dried—this means waiting at least 24 hours between the time the last piece of tile is installed and the time you start to grout. The directions on most packages of grout say "Mix to the consistency of heavy cream," which is just about right. You'll need some sponges and rubber gloves to work the grout in, and more sponges and some old towels to clean it up. Grout a small patch of tile first; setting times of grout vary according to the weather and conditions in the room in which you are working, as well as the type of tile you are installing. It is much more difficult to clean the grout off matte tiles than glossy tiles. Therefore with matte tiles you should do only a small area at a time, wiping the excess grout off fairly soon after you lay it down. With glossy tiles you can grout a much larger area at once and come back later and rub off the excess without any problems. If grout is allowed to sit on the matte tiles too long, it may become difficult or even impossible to get the excess off without leaving a heavy film. A light solution of muriatic acid can be used to remove excess dried grout the next day, but with its use you run the risk of taking too much grout out from between the tiles. If you have problems or have left any gaps in your grouting, repeat the whole process to build up a good layer of grout between the tiles. Grout should be allowed to cure for two or three days after installation and the area should be wet down with a spray-mist bottle (such as you might use for plants) each of the two or three days before you use the tub. These tiling techniques apply to all other tiling situations, such as stall showers, conventional tub installations, and even sink tops.

PLANT ENVIRONMENTS AND LIGHTING EFFECTS

One of the most satisfying final steps in setting up a loft for living is the installation of an area (or areas) where house plants can be grown and cultivated. If your loft is in a very sunny area, you won't have to do anything very special, except to acquire the plants and arrange them. Many lofts do not have good natural light, or have light for only a few hours or minutes a day. Until recently, you would have been quite restricted in the kinds of vegetation you could keep alive under

Large factory-sized windows contribute to the lushness of this indoor city garden.

these marginal lighting conditions. But with the advent of a variety of artificial lighting techniques, almost any space can be turned into a greenhouse.

If you are thinking of creating an environment for indoor plants, plan on having some form of artificial lighting and, if your budget allows for it, a system for properly humidifying the plants during the winter, such as a waterfall or steam-heated pebble tray.

For lighting, use either fluorescent or incandescent bulbs or a mixture of both. Plant experts claim that a mixture of the two different sources is ideal; but with the high cost of electricity you may want to stick to fluorescent lights, because they are much cheaper to run than incandescent (regular screw-base) bulbs. A bank of half a dozen 8-foot fluorescent tubes will give you enough light to keep a whole garden going strong. Ideally, you should augment these with a few clusters of special

Plants grown next to fire-escape door
can be augmented by fluorescent lights from above.

"plant lights" (available in 75- and 150-watt sizes). Supposedly the incandescent bulbs contribute to the flowering and greenness of plants, but we've got a whole bunch of trees and plants living quite successfully in a pure fluorescent environment. Individual conditions vary. You will probably have to experiment a bit to come up with a solution that suits you best. Under optimum conditions, a bank of fluorescent lights with a few incandescent light tracks is a good starting point.

Plant lights, fluorescent or incandescent, must remain on for at least 12 hours a day. Many people find it easier to have their plant-light systems on a timer. You can conceal fluorescents from view by building a valance around them.

Humidifying plants is necessary in buildings with steam heat. If there is a steam radiator near your plant area, you can make a very good humidifier with a length of pipe and a galvanized metal pebble tray (see figure 20). If you are not sure about working with steam pipes, get a plumber in, as a mistake could be dangerous. First, disconnect the existing radiator. Either throw the old radiator out or store it in case you move out and want to put it back in. Get a suitable length of steam pipe and cap one end with an air-release valve; at the other end attach a steam union to the existing supply valve. The end of the pipe with the air-release valve should be supported with a bracket or brace, and a sheet-metal tray with legs should be placed over the length of pipe. The tray should be deep enough to accommodate a shallow layer of pebbles and perhaps an inch or two of water. When the steam comes on, the heated water will evaporate into the air and raise the humidity level so the plants will not get too dry.

Figure 20.
A steam outlet can be utilized to fabricate a humidifying device for plants.

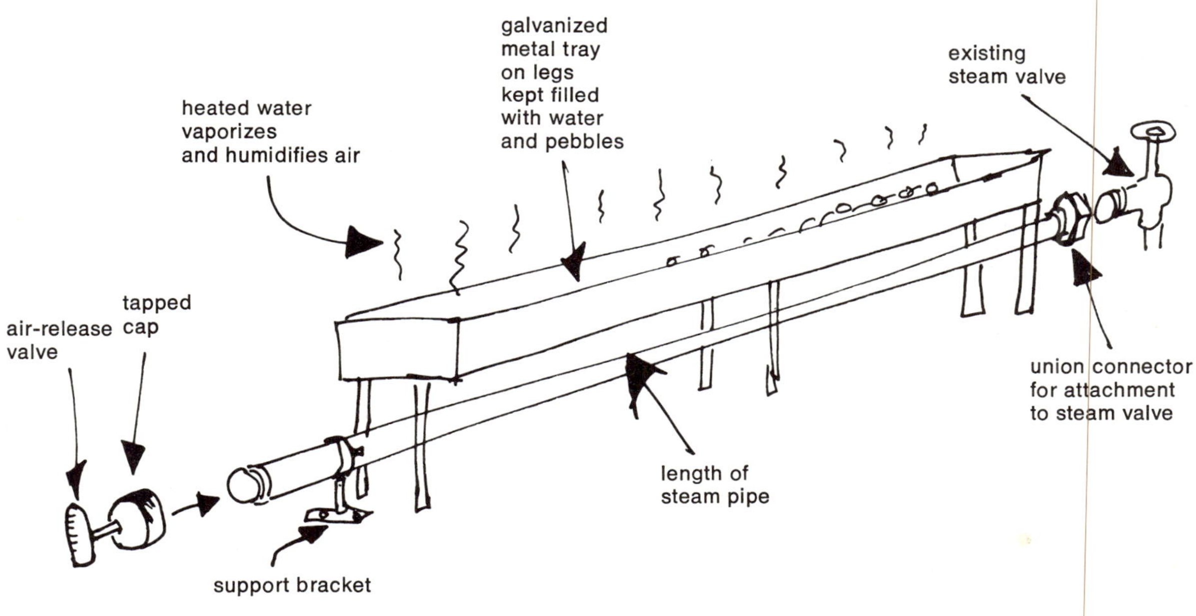

Waterfalls can also contribute to proper humidification of plant areas—and they make the air healthier for human inhabitants too. Some people create a whole bricked-in area for their plants to eliminate the need for individual drip pans for each plant, as well as to provide a natural setting for a waterfall. You can waterproof the area with fiberglass resin and matting, and then cover it with gravel chips for good drainage.

In addition to lighting systems for plants, lights can be used throughout your loft to create dramatic effects. Track lighting systems are very much in vogue for all types of living situations; people are installing them in apartments as well as in lofts and art galleries. Until a few years ago, it was very difficult to buy the tracks unless you were an architect or an interior decorator. Lately there has been a great proliferation of brands in track lighting systems, prices have been lowered through intense competition, and almost anyone can walk into an electrical supply house and order them. The great attraction of track lighting systems is flexibility, especially if you use a multicircuit track. When tracks are used in conjunction with dimmers and different styles of fixtures, almost any kind of effect can be created.

Colored theatrical spotlights and dichroic lamps will give ranges of color that can add an unusual dimension to loft living. Experiment with various combinations of white and colored lights to create the right mood for you.

We have presented just a few of the possibilities for turning a loft into an environment suited to you and your needs. By now you probably have a multitude of ideas, if you didn't have many or some when you began. The possibilities of what you can do in the open space of a loft are almost limitless. All you need is a little imagination, time, money, and, if you want to create your own tailor-made environment, the skills to do so.

In concluding, we can safely say that no matter how well prepared you think you may be, there's always some little surprise in store for you. You can't plan a renovation down to the last detail and expect it to go together without any hitches. You should be psychologically prepared for something to go wrong at least once or twice during the project. Hopefully, it won't be anything major or catastrophic—but sometimes pipes do spring leaks in the middle of the night and wood does get cut too short. Little things like that should not set you back. Press ahead! Do not give up the fight when building inspectors threaten you with eviction; with a little persistence, you'll be surprised how long you can survive. People who live in lofts are definitely a new breed; they are creating a whole new way of living and surviving in the mayhem and weirdness of the modern world. For an ever-growing number of people, loft living definitely is a better way.

Glossary

Glossary

anchor bolt: any of several kinds of bolts inserted and fixed in masonry

angle stop: a valve with a regular pipe thread on one end and a compression fitting on the other

black pipe: special threaded pipe used exclusively for gas lines; it is not suitable for water lines, as its chemical coating would give water an unpleasant taste and odor

BX: a pair of electrical wires clad in armored metal cable; used for wiring within walls

cast-iron pipe: pipe used only for drainage

ceiling plate: a horizontally laid piece of wood or metal used on a ceiling as an attachment device for studs

chalk line: a chalk-coated string used to mark out straight lines on floors or wallboard; the string is held taut against the surface and then snapped to transfer the chalk

composition board: a synthetic paneling material made by compressing wood fibers into sheets; available in the same thicknesses and sizes as plywood; also called Novoply

cut nails: special square-cut nails for driving into brick or concrete blocks

drainage pitch: the necessary slope drainage must observe over horizontal runs; proper pitch is ¼″ per foot of horizontal run

electro-metallic tubing (EMT): tubing used for surface-mounted electrical outlets, supply runs, and switches; also called rigid metal tubing

fiberglass insulation: spun-glass wool that is laid in walls or ceilings for insulating against heat and cold

flashing: a sheet of metal or asphaltic paper used to bridge the joints between a roof and chimney, or any vertical-to-horizontal plane, to keep water from entering

floor-through: a brownstone floor that is completely open, except for bathroom and kitchen

framing: the process of erecting a framework of wood or metal for the attachment of sheetrock or paneling

gaffer's tape: wide, fabric-backed tape with a good sticking capability; also called plumber's tape or duct tape

galvanized metal pipe: zinc-coated pipe, used in water supply and drainage lines, that must be threaded and assembled with pipe compound; a major disadvantage of this type of pipe is that it begins to rust and corrode upon installation and can occlude within five to ten years

gem box: a flush-mounted electrical outlet box that is used in hollow walls

hardened steel nails: *see* spiral masonry nails

hex-head bolts or screws: bolts or screws with hexagonal-shaped heads that can be installed with a ratchet wrench

high hats: recessed lighting fixtures; best installed in new ceilings, as they require the nailing of hanger bars to the ceiling beams or framework

joint compound: a premixed substance used for joining together sections of sheetrock to form a solid expanse of wall without apparent seams; used for spackling and plastering in general; also called ready-mix

joists: horizontal beams that frame a floor or ceiling; usually spaced every 12 or 16 inches apart across the length of the building to provide support

lag bolt: a heavy wood screw with a hex head or square head used for fastening two pieces of wood together; also used in conjunction with lead anchors to fasten something to a brick or cement wall

lath: wood strips or wire mesh used to support wet-mix plaster on walls

load-bearing wall: a wall that actually plays a structural or supporting role in a building; a load-bearing wall cannot be removed without jeopardizing the structural integrity of the building, unless a supporting beam of the correct size is used to span the opening

mastic: a paste type of cement used to put down tiles

muriatic acid: a caustic solution used in treating exposed brick walls to remove excess mortar and reduce falling brick dust

oakum: a fibrous material used for filling in the space between cast-iron drainage fittings

penny (d.) nails: the standard unit of nail size; the larger the number, the bigger the nail

plumb: precisely vertical

plumb line: a string with a weight (or bob) on one end; used to check verticality

polyurethane: a plastic-based clear finish used to make wood surfaces tough and water and heat resistant

PVC (polyvinyl chloride) pipe: plastic pipe used for cold-water supplies or drainage; can be mated to cast-iron systems by means of an adapter

PVCP pipe: plastic pipe used solely for hot-water supplies

reduction point: any point where the diameter of pipe used for drainage or water supplies is reduced

riser: a pipe that brings water or steam from the basement to individual floors

Romex: a cable that usually consists of a pair of electrical wires embedded in plastic; used for wiring within walls

sandblasting: the use of sand and compressed air mixed under great pressure and sprayed onto surfaces with a hose to strip paint or clean excess mortar off exposed walls

sheetrock: ready-made sheets of gypsum plasterboard; also called wallboard

shim: a small piece of material, usually wood, used to tighten two connecting pieces of material that would otherwise be too loose

sleepers: pieces of wood laid horizontally on an existing floor to which new flooring is nailed

slip joints or fittings: rough-brass or chrome-plated joints or fittings used to hook up drainage from sinks, showers, and tubs

sole plate: a horizontally laid piece of wood or metal on which wall studs rest

spar varnish: a good-quality varnish that can be used to dress down exposed bricks or to finish wood; excellent for outdoor and marine use

speedy connector: a piece of flexible tubing with a flange on one end for connecting to the base of the mixing valve on a sink or the cold-water inlet for a toilet

spiral masonry nails: nails used for driving into brick or concrete

stud: a vertical support member of a wall; either wood or metal

subfloor: any material used as a flooring underlay, over which a finished floor is laid

sweat copper tubing: copper tubing used for hot- and cold-water supplies

toenailing: driving a nail diagonally so that it goes through one surface of wood into another, thus joining the two

toggle bolt: an anchor bolt with hinged wings, ordinarily held open by a spring, that is pushed through a hole of a hollow wall or masonry; once the wings have opened, the bolt is screwed tight against the surface

toilet flange: a round brass or plastic ring that is placed on the floor to mate up the toilet to its drain

tongue and groove: milled lumber having a groove on one edge and a matching tongue on the other, so that successive pieces fit together

Index

Index